DREAMING THE BIOSPHERE

THE THEATER OF ALL POSSIBILITIES

REBECCA REIDER

Printed in the United States of America
14 13 12 11 10 09 1 2 3 4 5 6

LIBRARY OF CONGRESS CATALOGING-IN-PUBLICATION DATA

Library of Congress Cataloging-in-Publication Data

Reider, Rebecca.
Dreaming the biosphere : the theater of all possibilities / Rebecca Reider.
p. cm.
Includes bibliographical references and index.
ISBN 978-0-8263-4673-5 (cloth : alk. paper)
1. Biosphere 2 (Project) 2. Ecology—Research. I. Title.
QH541.2.R45 2009
577.072'079175—dc22
2009015586

Book design and type composition by Melissa Tandysh
Composed in 10.5/14 ScalaOT • Display type is Cronos Pro

CONTENTS

ACKNOWLEDGMENTS

Many, many people contributed to this book by answering my questions and telling me their stories about Biosphere 2—biospherians, scientists, builders, and others. Thanks to all of you for your time and energy, and for your willingness to share your experiences in the hopes that the rest of us might learn from them.

Three wonderful teachers helped me to develop this project at different stages as both I and the book grew up. Tony Burgess was a true mentor from when I first met him at Biosphere 2 in 1999. He shared inspiration and guided me to ask the big questions about Biosphere 2 and our own biosphere. Sheila Jasanoff at Harvard advised me on my undergraduate History of Science thesis on Biosphere 2. She contributed wise insight and astute questions to send my thinking in new directions. Fred Strebeigh at Yale introduced me to a wealth of writer's tools to transform this story into a finished book, and his enthusiasm was contagious. The book has benefited greatly from all of three of them.

Other friendly readers also made many helpful suggestions on the draft: Willoughby Anderson, Scott Slovic, Maura Leahy, David Spanagel, and my Environmental Writing classmates at the Yale School of Forestry and Environmental Studies. Thanks especially to Scott Slovic for helping me steer this book toward publication. People from Biosphere 2 who read the drafts included Tony Burgess and a few of the team still at Synergia Ranch: John Allen, Bill Dempster, Mark Nelson, and Deborah Snyder. Deborah and Mark made a truly dedicated effort at giving me detailed feedback, and the book is much stronger for it. However, this is not to say that any of these people fully endorses what is presented in this book.

I want to recognize the kindness of Roy Walford, who gave me support, encouragement, and a refreshing openness to getting the story told. I also thank Barry Osmond, who saw the historic importance of Biosphere 2 and opened research opportunities to me during his time as Biosphere 2's president.

Many people generously lent their photographs for the book: Lisa Walford and the Roy L. Walford Living Trust, Mark Geil, Peter Menzel, and Deborah Snyder of Synergetic Press. Thanks as well to Linda Leigh for her excellent help in archiving and sharing Roy's many photos.

The Mesa Refuge provided an amazing haven in which to recraft the book. My deepest thanks for that gift.

I have appreciated the work of the University of New Mexico Press staff in creating this book and thank editor in chief Clark Whitehorn and everyone there who have made the road to publication a smooth and enjoyable one. Jill Root's thoughtful copyediting provided just the right finishing touches.

I would like to express my huge gratitude to everyone mentioned here—as well as to the many wonderful people in my life, for keeping me going and sharing the journey. Big thanks to my family, including all who came before me, who in so many ways made this all possible.

DRAMATIS PERSONAE

Listing every person who ever contributed to the Biosphere 2 project would be a huge task. I provide this partial list to help the reader keep track of characters mentioned in this book. Those with a star by their names were interviewed for this book. Quotations in the text come from these interviews, unless noted otherwise.

THE STARS: BIOSPHERIANS, MISSION ONE (1991–93)

* Abigail Alling ("Gaie"), marine systems manager
* Linda Leigh, terrestrial wilderness manager
* Taber MacCallum, analytic chemist
* Mark Nelson, wastewater gardens manager, Institute of Ecotechnics director
* Jane Poynter ("Harlequin"), domestic animals manager
* Sally Silverstone ("Sierra"), farm manager and crew captain
* Mark Van Thillo ("Laser"), technical systems, cocaptain, and comanager of Biosphere 2 construction
* Roy Walford, medical officer

THE DIRECTORS

* John Allen, executive chairman and director of research and development
* Margret Augustine ("Firefly"), CEO and coarchitect

THE PRODUCER

Ed Bass, chairman of the board, Space Biospheres Ventures

THE COMPANY (OTHER KEY PLAYERS AT BIOSPHERE 2)

Norberto Alvarez-Romo, Mission Control director
* William Dempster, chief engineer
* Kathy Dyhr, director of public affairs
* Marie Harding ("Flash"), chief financial officer and board member
* Phil Hawes ("T.C."), coarchitect
* Kathelin Hoffman Gray ("Honey"), board member and director of the Theater of All Possibilities
* Silke Schneider ("Safari"), animal systems
Stephen Storm, plant tissue culture laboratory
* Bernd Zabel, comanager of construction, Mission Two biospherian

SET DESIGN

Walter Adey, ocean and marsh designer, The Smithsonian
* Tony Burgess, desert designer, University of Arizona/U.S. Geological Survey; professor, Columbia University Earth Semester
* Carl Hodges, agricultural systems designer, University of Arizona Environmental Research Laboratory
Ghillean Prance, rainforest designer, New York Botanical Garden
* Bob Scarborough, soil scientist
* Peter Warshall, savanna designer

SUPPORTING CAST

* Jeff Boggie, electrician
* Rod Carender, construction and maintenance
Jack Corliss, appointed as scientific director midway through Mission One
* Ben Epperson, early member of Synergia Ranch
* Randall Gibson, early chair of Institute of Ecotechnics
* Chris Helms, director of public affairs
* Ren Hinks, video artist
* Gary Hudman, computer systems designer
* Terrell Lamb, public relations for Ed Bass
* Tom Lovejoy, chair of Scientific Advisory Committee
* Steve Pitts ("Bear"), computer modeler hired in 1994

THE SEQUEL: BIOSPHERIANS, MISSION TWO (1994) (ROTATING CAST)

Norberto Alvarez-Romo
John Druitt
* Charlotte Godfrey
* Rodrigo Fernandez
Matthew Finn
* Tilak Mahato
Pascale Maslin
Matthew Smith
* Bernd Zabel

SCENE CHANGE

* Steve Bannon, interim CEO of Biosphere 2
* Chris Bannon, brother of Steve, chief of staff under Columbia University
* Martin Bowen, banker for Ed Bass
Bruno Marino, scientific director

SCENE CHANGE AGAIN

* Marlin Atkinson, marine scientist, University of Hawai'i
* Heidi Barnett, Biosphere 2 ocean researcher
* Wally Broecker, geochemist, Columbia University
* Michael Crow, vice provost, Columbia University
Bill Harris, president of Biosphere 2
Chris Langdon, marine scientist, Columbia University
* Guanghui Lin, lead on-site rainforest scientist
* Barry Osmond, Columbia's last president of Biosphere 2
* Adrian Southern, on-site rainforest researcher

SCENE CHANGE, YET AGAIN

* Travis Huxman, director, B2 Earthscience, University of Arizona

Doom! Doom!
You have destroyed
A beautiful world
With relentless hand,
Hurled it in ruins,
A demigod in despair!
We carry its scattered fragments
Into the void
Mourning
Beauty smashed beyond repair.
Magician,
Mightiest of men,
Raise your world
More splendid than before,
From your heart's blood
Build it up again!
Create a new cycle
For the splendors of sense to adorn;
You'll hear life
Chant a new and fresher song.

—Goethe's *Faust*,
adapted by the Theater of All Possibilities

PROLOGUE

ORACLE, ARIZONA, 2001

THE AIRLOCK DOOR CLANGS SHUT. AS A RESEARCHER SLIDES THE RED METAL handle down into place to seal off the outside world, our first sensation inside Biosphere 2 is overwhelming and immediate: the thick, heavy, humid greenness of the air. Even in the little entry room of cracked and mossy cement, it engulfs, dense and muddy with smells of vegetation and soil, life and death.

A wood plank path leads over ferns and mud puddles into the rainforest, where shocks of bright green banana leaves press against glass walls, battling toward the sun. Ants scurry in columns over branches and vines. The atmosphere hangs hot and moist; clothing dampens, hair wilts. And in some spots barely perceptible, in others roaring, drones

PHOTO BY MARK GEIL.

the noise: a relentless machine hum. An immense round fan rotates and whirs. Drooping fronds lilt upward as the fan ticks around, shifting air in the same pattern, hour after hour, with lockstep regularity. Aside from this blowing, however, the air is still—too still. In the absence of gusting winds that would force them to grow strong, young tree trunks sag, curving back down toward the ground. High above, yanking the trunks back up, ropes and pulleys crisscross the ceiling. Thin white steel beams etch a geometry of triangles against the blue sky beyond.

Out to the south of this rainforest, the glass roof soars over a woody grassland and rocky cliffs; below the rock face lies a placid ocean, and beyond it the tangled mats of a mangrove swamp. Suddenly, from somewhere in the marsh, a loud whoosh echoes—the wave machine pushing a perfectly formed ripple across the ocean's surface. The swell is perfectly calibrated, gentle enough not to disturb the gravelly beach, but strong enough to stir nutrients toward little nubbins of corals that grow in the shallows, on caged metal platforms. For this is not simply a garden in a greenhouse; it is, perhaps, the most highly engineered wilderness in the world.

Underneath this tropical paradise, the machinery of the Biosphere churns. Dim cement basement passageways weave beneath the wilderness, a sinuous network of white plastic and metal tubes streaming water and electricity to make the Biosphere go, each trafficking its input to just the right spot to emerge as rain or wind. Sounds and temperatures change at every turn in the maze. In one subterranean corridor, wave pumps roar; in the next room water methodically swishes and pours through cleansing trays filled with brown algae. Hulking cylindrical tanks hold various stages of the water cycle: "condensate storage," "wilderness rain." From within the subterranean maze, a long white tunnel, whipped by a cold rush of wind, leads to the inside of the Biosphere's "lung"—a huge domed room under a metal ceiling that subtly rises as air heats up and expands, so that the wilderness's glass walls will not pop; with each inflation and deflation, the Biosphere mechanically breathes.

A day in Biosphere 2 leaves the body damp and sticky, the brain buzzing from hours under glass that dulls the sun's rays but traps its heat. As the airlock door slams again, the air outside feels thin and spare, the mountains far away, the mesquite desert basin and dry riverbeds a Martian landscape. Imagine emerging after two years.

Only eight people will ever know what that final emergence feels like. But a river of RVs and family sedans bearing curious onlookers still makes the winding pilgrimage to this desert site thirty miles north of Tucson every day, searching for some clue to the mysterious rise and fall of one of the world's most biodiverse, lush wildernesses—and one of the most artificial. In southern Arizona's high desert grassland, on a low rise above a wide basin plain, sits what looks like a space age castle fallen out of the vast sky. At the foot of the craggy back slopes of the Santa Catalina Mountains, Biosphere 2's glass pyramids and arches gleam in the sun. The visitors drive up to the gate, park, pay their fare at the ticket window. They circle round the huge glass structure, peering through vegetation grown up against the windows as if searching for some clue of what once transpired inside. But neither the pictures snapped, nor three gift shops, nor little written signs scattered around the buildings, ever fully explain who would build such a place, who would fight for the chance to lock themselves inside for two years, and why—and where they all went.

In 1987, the popular science magazine *Discover* called Biosphere 2 "the most exciting venture to be undertaken in the U.S. since President Kennedy launched us toward the moon." The giant sealed greenhouse would officially be called "Biosphere *Two*," explained numerous newspaper articles, because the Earth itself was "Biosphere 1," and this megaterrarium aimed to be, as one journalist put it, "if not the discovery of a New World, at least the making of one." Media reports cooed over the "Planet in a Bottle," "Greenhouse Ark," and a slew of other mixed metaphors. The Biosphere's builders were widely quoted proclaiming their enormous and varied aims: to prove that humans, technology, and wild nature could coexist harmoniously under one roof; to make bold scientific discoveries about the workings of the Earth; and to create a prototype for ecological space colonies to settle other planets. Yet only a decade later, the fanfare was long gone; in 1999, *Time* magazine crowned Biosphere 2 one of the "50 Worst Ideas of the Twentieth Century."

As the project went from eagerly watched utopian venture to international laughingstock, journalists seized on Biosphere 2's weirdness—its appearance as a sort of tropical spaceship, its grandiose $150 million construction cost, and its creators' equally grandiose determination that their project would change the future of the universe. What few of the pundits recognized, however, amid the trappings of science fiction come to life—and what most of the tourists gawking with their faces pressed against the glass still fail to see as well—was despite all of Biosphere 2's weirdness, just how *normal* it might be, and how much its offbeat story was, in some sense, their own. Biosphere 2's founders packed their greenhouse world with more than 3,800 carefully listed plant, animal, and insect species, and tracked biological and chemical changes through countless scientific studies, trying to make sense of the interactions of organisms, soil, water, and air. Yet in the end, their miniature world became more of an unintended experiment in the behaviors of one particular species: *Homo sapiens*. "When you create a new world, you end up with all the problems in the world," one of its creators once reflected. The miniature world of Biosphere 2 was designed to condense the interactions of Biosphere 1's life forms, air, and water in a compact space. Likewise, human relations, and relationships between humans and the rest of nature, became intensely concentrated and displayed under its magnifying glass.

Peering through that magnifying glass, peeling back the layers of overgrown banana leaves and participants' various versions of history, this book asks two sets of questions. First, what happened to make that glass eco-castle so suddenly rise, and then suddenly fall empty? But second, what might the story of Biosphere 2 tell us about its namesake, Biosphere 1? And beneath that last question lurks another: what would it mean for people to truly get along, with each other and with all the other life forms on the planet?

Today the wool-carpeted halls of Biosphere 2's Human Habitat area are eerily quiet. A few researchers in relaxed work clothes walk between offices under the cavernous ceilings; laughing college students, visiting for the semester, bounce through on their way to the computer lab; behind glass doors, a handful of technicians checks computer monitors in the command room. Down the hall, the huge state-of-the-art kitchen gleams white, metallic, and empty. A few jars of dried beans remain on the counter, but just as a display for the tourists. Eight empty chairs still sit around the cold granite-slab dining table. And finally, high above, up a spiral staircase, shines the Biosphere's crowning white lookout tower, its windows commanding views of the desert arroyos, sunsets, and mountain ranges back on Earth. There, aloft in a carpeted library off limits to tourists, far above the sweaty din of the ecosystems below, an odd assortment of books still lines the shelves: collections of plays by great modern playwrights, ecology textbooks, legends of ancient civilizations, the Vedas, the Upanishads, literature on space colonies, how-to manuals such as *How to Grow More Vegetables Than You Ever Thought Possible on Less Land Than You Can Imagine*. Together, silently, the pile of books records a saga—a story of grand aspirations to understand and encapsulate the world. And tucked here and there amid all the books, other relics remain: a few slim volumes of play scripts from an acting company called the Theater of All Possibilities.

I first arrived at Biosphere 2 as a college student in the winter of 1999. The last human "biospherians" had left their sealed eco-home nearly five years before, and Columbia University was now managing the place, attempting to turn the glass spaceship into a climate change research facility and campus for environmental studies. Seventy-two undergraduates from schools around the country showed up that January to participate in Columbia's "Earth Semester" program. On this strange new college campus with only one lecture hall, on a clear, dark night with a starry desert sky opening up, Biosphere 2's latest president, a tall, well-dressed, and well-spoken former National Science Foundation administrator named Bill Harris, rose to welcome our new class. "At Biosphere 2, we're not looking to go to Mars anymore," he told us pointedly. "We are here to become better stewards of the Earth."

I did not realize at the time that this was one of the few official references to the facility's storied past that I would hear in my time at Columbia's Biosphere 2 Center. Signs around campus proclaimed "Research Today for Tomorrow" and described Columbia's own carbon dioxide research inside the glass structure, but scarcely mentioned the eight human residents of Biosphere 2, or the ideas that had given birth to the project in the first place.

I had to drive an hour to the University of Arizona library in Tucson just to find a copy of *Space Biospheres*, the book in which Biosphere 2's creators first presented their project to the world in 1986. In the slim volume, its authors, John Allen and Mark Nelson, concluded that whether by environmental collapse, nuclear holocaust, or the far-off death of the sun, eventually "Biosphere I must disappear . . . unless it can participate in sending forth offspring biospheres." Therefore, they reasoned, in order to enable life to continue in the universe beyond Earth's "localized planetary lifespan," humans must "assist the biosphere to evolve off planet Earth." By creating new biospheres, people could give life a permanent home on other planets, and eventually throughout the universe, Allen and Nelson argued. Biosphere 2 would be their first step toward doing this.

The little book's language was grand, full of esoteric flourishes and cosmological imperatives, but hardly seemed dangerous. But a few months later, when one of the authors, former biospherian Mark Nelson, came to visit Biosphere 2, it was only with the express permission of the Columbia administrators that he was allowed to set foot on campus at all. Under no circumstances, I heard from a staff member, would he be allowed to meet with students. Since major universities usually at least pay lip service to open inquiry into history, I was surprised that an institution like Columbia would have so much anxiety about the recent past. Someone seemed to want the world to forget that anything at all out of the ordinary had happened at this place—even as this enormous glass castle sat right in front of us, with its very name, Biosphere 2, inescapably recalling that someone here had believed that other worlds were possible.

I encountered the same perplexing pattern of uneasiness and resistance when, as a visiting student at Biosphere 2, I decided to start researching the place's story in order to write my undergraduate thesis in the History of Science back at Harvard. I met first with Tony Burgess, a biologist who designed the Biosphere 2 desert ecosystem. He was one of the few remaining staff members who had been there since the beginning. Marked as a gentle rebel by his bushy red beard and ever-present suspenders and field vest, Tony was one of the last human relics of the campus's early dreaming days. A devoted naturalist who somehow grew up in Fort Worth, Texas, with a deep fondness for desert plants, he had lived through the rise and fall of the 1960s, served in Vietnam, yet survived with his idealism intact—and he continued to draw in young college students seeking inspiration. From behind a desk piled with mountains of papers in his hivelike little office, drawling in an expressive, bemused voice, Tony loved to tell stories.

At first Tony seemed enthusiastic about advising me, happily rattling off tales about the cast of characters with whom he had built Biosphere 2. Yet a few weeks later, suddenly, I received an out-of-character, bureaucratic-toned email from him stating that my research proposal had been deemed "inappropriate" and that the history of Biosphere 2 was not a suitable subject for study. The email was CC'd to the project's top administrators. Tony told me later, in private, that he had been rebuked for agreeing to advise my project, but would not tell me by whom. When I asked him repeatedly why anyone could have so much at stake in an undergraduate research project, he would only gaze at me through his glasses and say, "You're more naïve than I thought you were."

Of course the Biosphere's administration did not have much control over whom students could visit the following summer.

SANTA FE, NEW MEXICO, 1999

Synergia Ranch is a collection of low-roofed adobe buildings set on hard, flat, gravelly earth, dotted with dried-up arroyos and juniper bushes, up a long dirt driveway off the road to Santa Fe. Abigail Alling, one of the four former residents of Biosphere 2 who still make their home at the ranch at least part of the year, is standing there waiting outside for me at the appointed time. She greets me with a warm smile, introducing herself. She briskly shows me around the cluster of apartments and offices, and through the large communal kitchen and dining room furnished with thick, hand-hewn tables and chairs from the 1970s. As we walk, she points out the high geodesic-domed theater building once constructed by the hands of the actors of the Theater of All Possibilities. Finally she brings me across a gravel courtyard to an office, to meet the man from whose mind it all began.

Rising from behind his desk, John Allen is tall, a large strong-looking man for seventy years old. Wavy gray hair lies combed across his head; he wears a button-down Western-style shirt and a ready smile. Yes, yes, come in, he urges, sit down at the long table. He is overjoyed to be hosting a History of Science student, he says, his broad face lighting up, particularly one from his old alma mater, Harvard; he winks, and declares the occasion fit for a Harvard History of Science tutorial.

The professor rises to the blackboard while the student opens a notebook full of questions; but before his interviewer can utter a word, he dives in: "Science is a series of conventions," he declares. "CAMPAIGN OF SCIENCE," he scrawls on the board. "IGNORANCE OF PREDICTABLE RECURRENCES OF PHENOMENA." Ptolemaic astronomy worked for sailing navigation, he explains; Newtonian physics will get us to Mars; Einstein is necessary to get us to the stars. Science is ever-expanding. He begins to scrawl names and arrows on the board, connecting them all in a dizzying history: Acadèmia de Lynxa, Galileo, Hooke, Newton, Linnaeus, Tuscany, Rome, Royal Society, Physics, Botany, Zoology. He is weaving an intricate picture of the history of European science at warp speed. The culmination of the physical sciences, he explains, was in astronautics; its emblem was the cyclotron. After several minutes of this, just as his observer tries once again to get a question in between breaths, he consents to mention the topic they are supposedly meeting about: "Biosphere 2 is the cyclotron of the life sciences."

His history lesson skirts all those he considers relevant: Russian geochemist Vladimir Vernadsky with his concept of the biosphere as a "cosmic phenomenon and geological force"; Harvard biologist E. O. Wilson and his idea of the "naturalist trance" as a mode for understanding and observing nature; another idiosyncratic Harvardian, the ethnobotanist Richard Evans Schultes, who lived in the Amazon jungle for more than a decade studying native people's uses of plants. Somehow he gets from thermodynamics to the Cold War and McCarthyism, and finally launches into his own story: from organizing in Chicago labor unions, to the sixties in San Francisco, to building this ranch with a few committed friends, to the long list of organizations and projects they started together: the Theater of All Possibilities experimental actors' troupe, the think tank they called the Institute of Ecotechnics, a ship that sailed around the world, a farm, a rainforest sawmill, a cultural center, a Biosphere. . . . He is maddeningly hard to follow (and refuses to be tape-recorded—the source only of "verbal diarrhea," he insists), impossible to interrupt, and totally fascinating.

That night, after a delicious dinner, wine, and ice cream with a group of the former biospherians, Allen is in impeccable form. He has learned that I am a fan of another of his offbeat heroes, the Beat junky-poet William S. Burroughs; and after regaling us all with stories about visits by Burroughs and acid guru Timothy Leary to Biosphere 2, he opens the latest Burroughs anthology and begins to read, in a fine, wry impersonation of Burroughs's dry voice, one of his favorite stories. All sit rapt around him on couches and cushions, giggling in delight.

It seems to me, however, that we still have not talked enough about the details of Biosphere 2; I request, through one of the women on the ranch, another interview. It is granted for the following Monday afternoon. This time we sit down facing each other across the long thin table in his study. But Allen's tone is instantly belligerent. He refers to the letter I wrote him months ago originally asking for an interview; why did I ask to meet "you and your group"? Was I implying they are some kind of cult? "Groupie groups" don't make history, he informs me angrily; *organizations* do. His fear of being labeled seems logical for someone who has been widely attacked in popular media; yet it approaches obsession, as he rails at me for half an hour without stopping for breath. Finally he pauses. "Alright, ask me your hardball questions." I realize anything controversial about the past is off limits, and venture instead a philosophical question about the scientific discipline of "biospherics." Instantly his face melts; he apologizes profusely, and with tears in his eyes, begins to sketch on paper, with building speed and excitement, the trajectory of human evolution as he sees it.

I had heard plenty of Johnny stories before I got to Synergia Ranch. Everyone who has met him has one. Both his friends and detractors had told me of his incredible ability to read and master information and to lecture on any set of ideas as if they were all his own—he is rumored to have collected highest honors at Harvard Business School while scarcely having to study. Furthermore, he is one of those people about whom it is impossible to be neutral. Some speak of him as a respected role model, others as a megalomaniac, others as a sort of poetic tragic hero. In one of the more ridiculous exaggerations, the *Village Voice* once called John Allen "more the Jim Jones than the Johnny Appleseed of the ecology movement."

For me, however, the most confounding thing about meeting the man, and the handful of Biosphere 2 builders and inhabitants who still live and work together at Synergia Ranch, was not, in the end, how strange they were—but how much, despite uncomfortable moments, I liked them. All along my trail through the past of Biosphere 2, on a road trip from California to the glass world itself to the northern New Mexico desert ranch where it all began, as I tracked down and interviewed those who designed, built, and lived inside Biosphere 2, I encountered engaging, intelligent, funny, lively, motivated people. Committed and creative, they seemed unafraid to try to change the world; they were driven by a love of life, a deep concern with environmental destruction, a determination to develop themselves and contribute to life on Earth (and possibly beyond). They had believed in a common goal, this alternate planet under glass, and had dedicated themselves to building it—even though things had not turned out as they planned.

And as much as I felt like a stranger at Synergia Ranch, after a few days of leafing through archives, eating, gardening, talking, and joking around together, I felt like part of the family too—and began to wonder, if I had arrived there thirty years earlier, whether I might have joined in. I knew about John Allen's bag of tricks—his storied skills as an actor, his ability to transform his own persona to appeal to the thoughts and desires of the person in front of him (such as his playacting of a Harvard tutorial and his drawing on my admiration of William Burroughs). And yet, even if the science was inseparable from the theater, among this group of people I began to see how contagious ideas and excitement could be—particularly the notion that indeed any dream represents a real possibility. As I met the creators and stewards of Biosphere 2, I came to realize that the performance they put on together as the Theater of All Possibilities, on stage and in life, was not merely an off-the-wall farce. It was a microcosm of a grander drama about humans and ecology at the end of the twentieth century. I discovered that the story was not just about one group of dreamers, but of a culture's desperate quest to transform the destructive relationship between humans and the rest of nature; a quest to create a more beautiful and perfect world, and to create the human social forms up to the task.

ACT I

The RV *Heraclitus,* one of the earliest major creations of the core team who would go on to build Biosphere 2.

PHOTO BY MARIE HARDING. SYNERGETIC PRESS.

SEEDS

THIS RANCH IN NORTHERN NEW MEXICO WAS WHERE BIOSPHERE 2'S founders ended up. It was also where their world had once begun.

In the shadow of bumpy peaks, I walked out into the desert at dusk on the gravelly rangeland. In gullies I noticed little mounds of debris—piles of bricks and rocks stuck in shallow dirt channels, where water would probably gush in a rainstorm. When I asked John Allen about them, he smiled. Those humble little dams, he said, represented the long-ago beginning of "ecotechnics," the name he and his fellow Biosphere-builders gave to their work with the earth. They had experimented with channeling storm runoff to reduce erosion across the ranch landscape thirty years ago. One morning during my visit, as a few of us took a break from planting peas in the soft brown garden soil, one former biospherian pointed out to me a sort of crumbling cement bunker dug down into the ground nearby. This half-open "grow hole," as she called it, didn't look like it had much in common with Biosphere 2. But three decades ago, it had been the site of one of the group's first experiments in growing plants in a controlled environment.

The landscape of Synergia Ranch was understated. Just a cluster of smooth adobe dwellings on an open plain, a geodesic dome theater, and an expansive dry orchard full of gnarled fruit trees. It sat amid the quiet beauty of the high desert dotted with scrubby plants. But even though there were few physical clues to show it, in the early 1970s the seeds of Biosphere 2 had been planted—if not physically, at least conceptually—on this windswept homestead. Here idealistic and passionate young people had come together out of the chaos of the 1960s, carrying their despair, but still harboring dreams of a new, better civilization.

Synergia Ranch was a special place, an historic place, not so much for its mud walls as for what those walls stood for. Those hand-built dwellings had risen from the ground at an amazing and brief moment in American history—a moment when large numbers of young people actually believed that together they could create a better world. On this ranch, many of them—"refugees" from society, as John Allen called them—began that creation process together, from the ground up. "It was just absolutely desolate, it was ecologically desolate," recalled Ben Epperson, an early Synergia resident. "So everybody got to work, and started planting." The ranch residents planted eight hundred fruit trees, and other trees for shade. They piled up those little "ecotechnic" dams of rubble to slow down rainwater as it flowed across the land. Their first construction projects began humbly too. Learning the traditional adobe techniques native to the Southwest, they built their new homes out of the most readily available resource: dirt.

Years later, when the glass-and-steel pyramids of Biosphere 2 at last rose into the sky in the 1980s, media reports would bring up the founders' past shared ranch life as incongruous, or even scandalous. Vehement critics would try to discredit Biosphere 2 by questioning the fact that a dirt-poor, back-to-the-land commune had eventually transformed itself into a corporation with a mammoth $250 million science project. But what I found most remarkable about Synergia Ranch was not the transformation from humble beginnings to huge corporate science experiment, but that the ranch embodied such grand ambitions from the beginning. One early member, a young experimental architect named Phil Hawes (later architect of Biosphere 2), recalled that in the ranch's first year, the dashing ringleader John Allen "had put up a piece of paper on the board that said something like, if we work correctly or do things correctly—he had kind of an outline—in fourteen years we should be able to do a world-class ecological project." That was in 1970, Hawes said. And, in 1984—exactly fourteen years later—"we got the land" for Biosphere 2.

The place's name embodied the scope of the dream. The word *synergia* invoked the concept of "synergy" described by the inventor-philosopher Buckminster Fuller, one of the ranch members' many counterculture intellectual heroes. "Synergy," as Fuller described in his books, meant that the whole was not predictable from the sum of its parts. In his *Operating Manual for Spaceship Earth,* Fuller gave one of his favorite examples of synergy: stainless steel. This alloy, made up of various metals blended together, possessed properties that no single one of its constituent metals possessed on its own. John Allen, who had studied metallurgy, understood the example well: "you can get stainless steel by putting together metals from whose separate properties you could never predict this amazing corrosion-resistant, beautiful result," he wrote.[1] Likewise, the ingredients in the alloy of Synergia Ranch would be diverse: ecology, art, theater, philosophy, meditation, hard work on the land, intense examinations of group dynamics, and young, energetic people. Out of these pieces, the synergians believed they could create a reality far greater than the sum of its parts. Their inner goal was nothing less than a new civilization, a "synergetic civilization," as they liked to call it. As I spoke to former ranch residents, their faces lit up warmly as they recalled the wild early days, their shared optimism, their sense that they were breaking new ground. What was the purpose of the ranch? I asked Marie Harding, who had been John Allen's wife at the time. "Synergetic civilization!" she told me without hesitating—then added, "Whatever that is!" and laughed.

At the beginning, that spirit of synergy was one of the group's few and important assets—that, acres of overgrazed pastureland, a worn-out stable, and a few members' savings accounts. The ranch members came from all over the country and beyond, showing up one by one as word of mouth spread. The new synergists were mostly young, intelligent, and unexcited by the life options offered by the rest of society at the end of the 1960s. And despite the small scale of their little ranch society, they came to believe fervently that there they could create real change—starting by creating a better miniature world right there among themselves. "We were only 8 or 10 persons who were not satisfied with how our lives were with respect to the world," wrote Phil Hawes, the young architect. "There was a painter, a French teacher, myself (being a graduate architect and town planner), a folk singer, a keeper of bees, a mining engineer, an anthropologist, a poet, and a cook. We were convinced that the world was not doing well, and we decided to change our lives, coalesce our wills and skills, and, by turning our attention to addressing questions of ecology and the environment, take steps to alter world history."[2]

In their simple lives and in their enormous aspirations, the synergians were simultaneously unique and totally typical. They were playing out a pattern common to their historical moment: the quest to create a better world, on a small scale, through the intensely shared work and life of a group of people. That social pattern itself was just one iteration in a recurring theme that had already replayed itself several times throughout Western, and particularly American, history: the repeated flowering, in concentrated bursts, of communal work and living experiments. At the end of the 1960s, in one explosion of such energy, hundreds of new communes were sprouting across the North American continent, first in cities and then all over the countryside, especially in the West. It was no coincidence that this movement occurred at the tail end of the 1960s. Many idealists and activists, after years of protest, felt frustrated with the lack of broad changes in society. Rather than give up on their ideals, they turned inward, to create new, more harmonious little societies for themselves. The establishment of rural communes peaked in the years 1969–71, according to commune historian Benjamin Zablocki.[3] Synergia Ranch began in 1969. The landscape of the American West attracted many of these communal pioneers. They were looking for a blank space on the map to create something new.

But there the story of Synergia diverged from the history of most contemporary communes, through a simple fact: it survived, even prospered. Most late-sixties communes boomed and busted, as members drifted together and then drifted away. As Ram Dass wrote in 1971, "Many spiritual seekers have joined or started communities . . . Often they have been disillusioned by these experiments because of disorder, economic instability, ego struggles, and mixed motives on the part of the participants."[4] Historian Yaacov Oved put it more succinctly: "The main feature of these communes of the 1960s was their transience."[5]

From the 1960s onward, activists for a host of causes have repeated like a mantra the famous words of anthropologist Margaret Mead: "Never doubt that a small group of thoughtful, committed people can change the world. Indeed, it is the only thing that ever has." But obviously, plenty of small, committed groups have failed to change the world as well; the short-lived communes of the 1960s are a prime example. So what does it take for a group of people to actually come together and create a functional new world? Why do people keep trying to do so?

And why is it so hard to achieve? These would become key questions at Biosphere 2; but these were also the questions that Synergia Ranch, and other intentional communities, had been living from the beginning.

Of course, Synergia Ranch's creation story depended on the combination of a particular handful of people and their specific personalities, backgrounds, and visions. But as the present and past synergians shared with me their memories, patterns emerged. Synergia's genesis was a story not just about community living, but about the human yearning for connection and meaningful life—about both the possibility and the difficulty of realizing that dream on the land.

The hardest part of uncovering the history of this commune, however, was that no one even wanted to call it a commune. "There was no commune in Santa Fe in the late '60s or '70s in which I was involved," John Allen insisted in a scathing letter to a newspaper in the 1990s. Synergia Ranch was a business based on "contractual relationships" between "individuals who wanted to experiment with an ecotechnical approach to setting up a business," he insisted.[6] Perhaps Synergia Ranch was the anticommune commune (though when I tried this phrase on John Allen, he told me I was speaking "Harvard drivel"). "He will not acknowledge that the ranch in Santa Fe was a 'commune' by anybody's standards," recalled Kathy Dyhr, who moved to Synergia in the late 1970s. "But on the other hand . . . we lived together twenty-four hours a day, and worked together." Even at the time of the ranch's founding, at the height of the commune era, that word was not uttered at Synergia. Laurence Veysey, an anthropologist researching the American communal living phenomenon, visited Synergia for five weeks in 1971. He described the ranch and its residents, using pseudonyms, in his book *The Communal Experience*, depicting John Allen as a dark prophet named "Ezra." As Veysey observed, "The word 'commune' is never heard around here, except as an unfavorable epithet, for it implies the disorder and lack of conscious planning and effort which this community strongly rejects."[7]

There were, of course, reasons for the resistance to the word. As John Allen himself told me years later: "groupie-groups" don't make history; "*organizations*" do. Though the ranch-dwellers lived in a corner of the mountains of rural New Mexico, running their minibusinesses, they already had planetary-sized aspirations. Laurence Veysey paraphrased the worldview that Allen put out to his friends in teachings and speeches:

> Western civilization isn't simply dying. It's dead. We are probing into its ruins to take whatever is useful for the building of the new civilization to replace it. This new civilization will be planetary. The whole earth will be our home. We are no longer Americans, or Westerners, even though as individuals we were once trained in that tradition. We will build a series of centers in various parts of the world to demonstrate the new way of life. This ranch is merely our first training ground.[8]

The ranch members took that directive to heart, said resident Phil Hawes. "In short, we planned to help pick up the pieces of the cultural shambles that appeared to be the wreckage of Western civilization. We planned to salvage the best from the past and press on to build a more acceptable and satisfactory future."[9]

"Every now and then people would call us hippies, but that really wasn't too accurate," agreed early resident Ben Epperson, describing the group. "At least 50 percent of them were upper-class dropouts who didn't want to drop out and just goof off, but they were sort of disillusioned with whatever their situation was." They were "young, bright, highly motivated people; they worked their asses off, and had Harvard Business School management direction by a sort of non-standard genius."

That "non-standard genius," of course, was John Allen. If the story of Biosphere 2 must start with the story of Synergia Ranch, then the story of Synergia has to start with Allen's story—for while the ranch was born of many people coming together in "synergy," it could not have been created without his impetus. In turn, the successes and tensions in the ranch would be linked to his own personality.

His friends called him Johnny. When he talked his whole body became animated, his mind spinning ideas, seemingly effortlessly. Even when I met him when he was seventy, his natural eye-twinkling charisma, a certain Western ease, still drew people to him. He considered himself a son of the Western frontier, born and raised in Oklahoma in the 1930s. From early in life he had showed leadership and intelligence—and restlessness. He was strong and quick-witted, a high school valedictorian and multisport varsity athlete. Beginning at the age of fourteen, his wanderlust kicked in; he roamed the West working on fruit orchards in the summers. After high school, he bounced between studies at four different universities before getting his bachelor's degree in engineering at the Colorado School of Mines, where he was also student body president. He then enrolled at Harvard Business School and graduated at the top of his class, with High Distinction. In between his various studies, he zigzagged back and forth across the country. He roamed the West, became a fruit farmer in Washington state, worked as a leftist labor organizer in the Chicago packing houses in the 1950s, joined the Army Corps of Engineers, made mining discoveries, and finally turned into a young businessman working on international mining development projects in Iran and the Ivory Coast. Along the way, he wrote poems and plays out of his experiences. "The guy's always full of ideas, of doing things," said his friend Ben Epperson, a fellow talkative self-styled philosopher who met Allen in Oklahoma in the 1940s. "He wanted to be a writer . . . I thought he was going to be the American Tolstoy. He turned out to be something somewhat different."

Throughout his adventures, John was also collecting ideas, stowing them in his mind for the future. He had a knack for memory and synthesis; friends recalled in awe how he could absorb a book, then give a lecture on it as if the ideas were his own. He was also collecting heroes and intellectual role models. At Harvard he watched the visionary Buckminster Fuller give a five-hour speech; the hundreds of people in the audience gradually faded away, leaving just a handful who "were hungrier for ideas than food," including John.[10] At Harvard Allen also first encountered the writings of W. R. Bion, a British psychologist who studied group behavior during World War II. Reading Bion, Allen became intrigued by the science of group dynamics, a subject that would become increasingly important in his own life. Later, as a businessman, Allen apprenticed himself to international entrepreneur David Lilienthal, the man credited with inventing the term "multinational corporation" in 1960.

But all these eye-opening experiences were not enough for him. John continued to roam, searching out minerals to mine on Lilienthal's international projects. And then, at the age of thirty-two, in 1963, he found himself sitting in an office building in New York City. One day he had an epiphany. He looked out at the sky and realized he could not even open his own window in his skyscraper office. He was trapped by his supposed success. At that moment, he saw a Yugoslavian freight ship heading out to sea, to North Africa. And he decided to go, as he later recalled:

> Those are all gorgeous windows in my shining skyscraper giving me million-dollar views of the world's splendid traffic, but I can't open them to smell the fresh air or hike barefoot along an open road or, worst of all, walk away from a pompous power holder. Get me outta here! . . .
>
> I knew what I had to do next—get on the next Yugoslav freighter headed East, to Tangiers. I had people to see and places to go where I had not the slightest idea what would happen next.[11]

The next phase in his true education began. He shipped out to North Africa on a Yugoslavian freight ship, headed for Morocco, the escape destination of choice for avant-garde American artists such as the Beat generation. He stayed there studying local art, hitchhiked across Africa, then hopped another freight ship on to India. He was coming to realize, as he later wrote, "Many of the masters I've met didn't graduate from any university of books, but they had all graduated from the university of life."[12] He longed to meet more of those masters; perhaps he longed to become one himself. He sat at the feet of Hindu karma yogis and Tibetan lamas, seeking new teachings. As Ben Epperson, who received postcards from his friend, put it, "He sort of realized in the East there were things that he didn't understand how they did; he wanted to find out."

In India, in Old Delhi, Allen also met Marie Harding. A long-haired, bright young American woman from a well-to-do East Coast family, she was, like him, making her way around the planet, "to get a picture of what's really going," in Allen's words. They met up again in Vietnam, in Saigon. Marie was working as a field nurse at a jungle hospital for tribal peoples in the mountains. John followed her there; she was "one of the most fearless and capable planetary explorers I have ever met," he later wrote.[13] They got married there in Vietnam.

In 1967, John and Marie moved to San Francisco together. Allen had traveled around the world for two years but was still hooked on exploring. He began cooking up plans to move to Australia to live with aborigines. But the explosive atmosphere in late-sixties San Francisco hooked him. As he put it, "Here were twenty thousand Americans setting up a cosmic continuity with their bodies and minds in Haight-Ashbury!" He "had always loved oil and mining boom towns," he wrote in his semi-autobiographical novel *Liberated Space*, an account of the San Francisco scene. "Now I watched the Haight, a boom town of experience." In that novel, a thinly fictionalized account of his own stories, he described a swirling of gurus and seekers, communal rooming houses, sex, grass and acid, strangers who instantaneously became intimate friends on the streets and in the parks, dance, theater, enthusiastic discourses about the nature of the universe. The young people of San Francisco "were creating something decisively new," Allen wrote. And, "'seeing' this was not enough for me. I determined to experience it all."[14]

Ben Epperson, who met up with his old Oklahoma friend John in San Francisco, agreed. Ben had been studying for a PhD in philosophy at USC, but by the mid-sixties, he said, "I was just burnt out" and "ran out of energy and money." Like John, Ben talked and talked with a lively, wry intellect, and seemed to feel it was natural to bring up everything from Wilhelm Reich to the French Revolution dictator Robespierre while discussing the history of Biosphere 2. Still, he had never seen or heard of anything like what was happening in San Francisco, he said. "Haight was this tremendous congregation of energies. And we got some of the best kids in the United States. I was born in '26 so I was about forty, and I'd been through bohemian and radical movements and everything before that . . . It had the greatest energy and the greatest creativity of any period I've ever seen."

In that vortex of creative energies, where anything seemed possible, a group of cocreators began to come together. "'Recruiting' would be the wrong term, but people were being attracted to the ideas and energy," recalled Ben. Waiting at a bus stop one day, John Allen met Kathelin Hoffman Gray, a willowy young woman with long dark hair and the beguiling air of a performer. She was studying dance with the famous teacher Ann Halprin at the time. They began talking. John described their meeting in his novel:

"I dance, but I want to do theater," she said.

"I travel, but I want to write theater."

We stopped to gaze into each other's eyes, trading energies and enigmas.

Allen knew that if they kept up the conversation, "she would be the first member of my karass, that circle of beings that recur throughout the generations all of whom it is necessary to meet and connect with for an individual to accomplish destiny in his lifetime. I knew myself well enough to know the discovery of my karass had to be my next step."[15] Indeed, the karass was coming together: John and the two women close to him, Marie and Kathelin, would one day sit together on the board of directors of Biosphere 2. Together, they and a small group of friends began rehearsing experimental plays. To support themselves, they opened a coffeehouse and restaurant on Sutter Street, which they named the Sign of the Fool, after the tarot card. In its bright front room they served sandwiches and light meals, and also held their first theater rehearsals. As the theater group grew to twenty people, they began having acting classes and rehearsals in an abandoned grocery store.

They were devoted to theater as a way of bridging their fantasies and ideas and real life. "I really didn't want to do traditional theater, and I didn't want to do just experimental theater. I wanted to do something that meant something for me, that had some kind of result, that was integrated with social change but that wasn't just propaganda," Kathelin recalled.

But John's frontier-seeking spirit persisted. Eventually the San Francisco high was wearing off. He and his friends felt ready for some new source of excitement. "Anybody can run an experiment in Haight-Ashbury," Allen recalled thinking, "so let's see if we can run one in Manhattan." Following his lead, the small theater troupe cleared out and took off to New York City. Pooling their money, they rented a house in Brooklyn and a theater studio in Manhattan. For their premiere they staged Sophocles' *Oedipus the King*—according to their own interpretation. "I don't really know Greek, but there were no good translations at the time so I rewrote the thing," Ben Epperson laughed. At its height, the actors' troupe grew to fourteen members. But theater was just one part of their intertwined lives. An informal school began to grow up around

John Allen. Soon the group's modest goal emerged: to draw on the best of world civilizations and create something new. Every week, friends gathered at night to hear John lecture on metaphysics, history, and world cultures. The guests came in costume according to the time period or theme of the evening. One week they were living in the Wild West, the next week in Elizabethan England or ancient Greece. Other times, Allen gave rambling, excited lectures describing his far-out visions of the future; his wife, Marie, recalled hearing him describe an undersea civilization, while another attendee, Randall Gibson, remembered him giving a talk called "Psychokinetic City."

Yet talking and acting on stage were still not enough. The restless ringleader was already eyeing his next challenge. It was to act in real life, bringing together all he and his friends had learned. And so, in late 1969, a bigger drama began. The actors and friends pulled up their roots and crossed the continent once again. This time they headed out to the high desert of northern New Mexico, outside of Santa Fe, to begin a new life together on the land.

The flat 165-acre parcel was little more than arid, scrubby pasture lying under brown mountains. It was badly eroded from years of overgrazing, with only sparse brushy vegetation on the gravelly soil. Marie Harding cashed in her family inheritance to make the $15,000 down payment on Synergia Ranch.

As word of mouth traveled, acquaintances and strangers from around the country began to show up. Randall Gibson, who arrived at Synergia from the East Coast on New Year's Day, 1970, described the early ranch atmosphere: "We had not defined anything except there was theater, there was philosophy, and people had to have an enterprise because we were broke." An upper-class but disillusioned graduate of the University of Pennsylvania, he decided to join in. Sometimes random people just walked up the dirt road with their belongings and asked for a place to stay. Anyone was welcomed as long as they would work. "Synergia Ranch had a policy, unless you were absolutely, totally weird, or schizophrenic—we got one schizophrenic, a couple of dope addicts, and so on and so forth; they were quietly ushered out—but if you were any kind of reasonable individual, like some Jesuits came and stayed, you could stay there for three days," Ben Epperson said. At the beginning, the only built structure was an old stable with a few big rooms. As newcomers arrived, they nailed up boards to make walls between the horse stalls so that each person could have his or her own little territory.

John Allen, the business school scholar, set up the ranch with a formal economic structure. Each resident paid $45 a month for room and board, worked four hours a day on projects around the property, and worked four hours a day on his or her own minibusiness. The ranch members' professions ranged from mechanics to artists; everyone worked out some kind of economic enterprise to support themselves. They paid 10 percent of their earnings to the ranch as well. Residents' fees went toward paying for the ranch land, buying food for communal dinners in the dining hall, and expanding the ranch to include a garden and an orchard; a barnyard full of pigs, cattle, and chickens; and more buildings to house the flow of new arrivals.

There was nothing easy about ranch life. "It never cracked up totally, which it almost did," said Ben, "because the money, for example, was always very, very, very tight for years—how were you going to pay your grocery bill? The hippie idea of being totally self-sufficient just was not possible . . . The food wasn't superb in the earlier years."

"Sometimes I call it the bad old days; sometimes it was a lot of really, really hard work," said Phil Hawes. He described long workdays, straightening nails outside by lantern light on freezing winter nights. One winter, he recalled, "it hit thirty below zero, and we didn't have heat in the rooms. So the pee froze in the pot. It never was an easy life, it was always very difficult, in the sense of what normal people, 'average' people think of as difficult."

However, many who showed up at Synergia—and particularly those who stayed on—were not necessarily seeking an easy life. The arrivals ranged from dissatisfied upper-class rebels to those from rougher working-class backgrounds. But most of the core group who stayed, as Allen put it, were a young "Ivy League establishment kind of set, but free thinkers."

The ranch members each had idiosyncratic backgrounds, but as they told me their life stories, a pattern emerged. They were all determined to live a more meaningful life, searching for something—though until they arrived, many did not know exactly what they were searching for. Often, in interviews, I noticed that the core members' voices seemed strained, a bit edgy, as they spoke about the days of Biosphere 2. But their faces almost always became excited, their tones nostalgic, as they recalled the ranch's idealistic beginnings. When they arrived in New Mexico, most had felt dissatisfied or powerless as lone individuals in society. In contrast, at Synergia, they lived a full life of work, theater, study; they felt the spirit of a real community; and they felt like whole people. That sense of personal and collective engagement, of the body, mind, and heart, was the first, and perhaps most important, ingredient in their new world.

The cast of characters gradually assembled. Phil Hawes, a talented, free-thinking young architect, moved to Synergia in 1970. Phil had trained with Frank Lloyd Wright at his famous Taliesin school. Like others drawn to the desert ranch, he was not content to pursue a normal career. He went by the nickname "T.C.," short for "Thundercloud," in honor of the thrill he got from flying an airplane in the updrafts of storms. He liked building with adobe. And in his spare time, he sketched fantastical turreted castles that he dreamed of building out of earth-friendly materials. He would one day draw blueprints for Biosphere 2.

Stephen Storm would one day run the plant tissue culture laboratory at Biosphere 2, culturing plants for space travel. In his twenties he had lived in Los Angeles until his frustration with society finally drove him away: "I just became basically dissatisfied with the way the world was being run, and at some point decided to withdraw my efforts from that system. And I said, 'I'm just not gonna work here any longer and pay taxes into this system that builds these damn freeways, and makes this Los Angeles look so awful,' and I just basically said, I dropped out."

He moved with his wife and children to Texas and started a landscaping business, enjoying the progressive "anything goes" scene in Austin, where he joined a communal dinner group with like-minded people. But still, he said, "I sort of worked alone and played together," and it wasn't enough. "I could see that what I was doing by myself would never get anywhere, it would never produce an alternative. And so I began to look for some way of creating or constructing an alternative to what I didn't like, because all I could see myself doing was sort of living around the fringes and complaining, and that just didn't suit my idea of a way to live."[16] Around 1970, on a trip to see friends in Santa Fe, Storm spent a weekend at Synergia Ranch. He went home and told his wife he wanted to move there.

Bernd Zabel would eventually manage construction of Biosphere 2, and would briefly live inside it. The blond-haired German had a long journey to get there, though. He spent a total of eleven years at university in Germany, studying engineering, then philosophy, then education. Finally he became an electrical engineering teacher. But the life of a civil servant seemed too set, too regular, he said. "Two years you're an apprentice teacher, and then you get your final exam . . . unless you do some crimes, you know exactly how you proceed. It's secure; it's a government job." At age thirty, Bernd traveled to America "as a scout to investigate alternative lifestyles, and was hoping to implement and create something like this with friends in Germany." He visited various communes around the country, but the energy level he found at Synergia inspired him the most. He went home to Germany to talk about recreating such a place with his own friends. But eventually he felt himself drawn back to New Mexico, where the synergians were already acting out bold ideas instead of just theorizing them: "In America it's just do it—you don't think much, you do it. In Germany I spent two months discussing until five o'clock in the morning the philosophy of this lifestyle, and realized that it probably never would happen." He had fallen in love, he said, with Synergia Ranch's efforts to "create a balanced man": "I was a German romantic and totally attracted to this Synergia lifestyle. It was not just a lazy hangout where people got drunk or stoned; quite opposite, it was really very active, from meditation, theater performances, to book readings and very hard physical work. It was a very fulfilling lifestyle that attracted me."

Kathy Dyhr, later Bernd's wife, would one day direct public relations for Biosphere 2. Though American-born, she too reached Synergia through an around-the-world quest. She also told a personal story of disillusionment and hope—disillusionment at the state of the world, and hope that somewhere, perhaps at Synergia, she could help create something better. Kathy had studied social work and nutrition at the University of Minnesota, then dropped out and started a halfway house for women, which she directed for five years. She felt compelled to act on her ideals of a better world, she said, so she traveled to Africa to become a volunteer nutritionist in Kenya and Nigeria. When she got there, however, she felt frustrated and powerless in the face of drought and famine:

"There wasn't a lot a nutritionist could do, since nutrition recommends good food, but there was almost nothing about how to get some if you don't have any. Of course the problem was environmental devastation." Kathy happened to meet the synergian architect Phil Hawes while he was traveling through Nairobi on one of his own adventures, and Phil told her about Synergia Ranch. When Kathy got back to the United States and made her way to New Mexico, the ranch hooked her too. "You could consider all sorts of ideas and how they applied to this and that and the other. But there wasn't a dogma," she recalled. "And to have an association of people who were based on the fulfillment of the highest potential of those individuals as well as the environment around them . . . to me that was a very interesting thing."

For many of Synergia's well-educated residents, the ranch's biggest draw was not just the flow of ideas, but the chance to actually *do* something with them. Mark Nelson, a future biospherian, had graduated from Dartmouth summa cum laude in philosophy. He was on track to become a doctor in a family of professionals. After graduation, though, he was working as a taxi driver in New York City, and he realized, quite simply, "I was an urban Ivy League–educated ignoramus, without any practical, real-world experience." Everywhere around him, in the late 1960s, he saw people experimenting, and he became possessed by the urge to "do something more fundamental." He headed west to Synergia and eagerly joined in the daily hard work of restoring the ranch land, improving the soil, and planting seeds.

Synergia was answering a hunger that young people were feeling across America. It was a hunger for meaning. *Modern Utopian,* a magazine for aspiring commune dwellers at the time, described the attraction of alternative communities in 1968: "All *viable* communities have a conscious intent and meaningful function. As opposed to the lack of meaningful direction in today's American society, a conscious sense of purpose is the actualizing force."[17] To help them realize that sense of purpose, the magazine offered its readers tips on everything from finding cheap land to bulk granola recipes. Historian Louis J. Kern described the mass movement toward communes at the time:

> Although today's communards are characterized by anomie and rootlessness, their alienation is that of a privileged class

> (upper middle, upper) . . . "the children of prosperity." Their revolt is grounded in disappointed expectations: American values, political and ethical, are hollow . . . man in a materialist, technocratic consumer culture has lost touch with his ability to feel and to respond spontaneously and, perhaps most importantly, with his desire, his need, for transcendence.[18]

There was a good reason that Synergia and so many communes were germinating at the same time. As the 1960s came to a close, the Vietnam War was still raging; meanwhile, the sense that total social revolution was just around the corner gradually faded. Around 1970, "[t]he riptide of the Revolution went out with the same force it had surged in with, the ferocious undertow proportionate to the onetime hopes," wrote cultural historian Todd Gitlin.[19] Radicals were left wondering why, after so many marches, sit-ins, human social experiments, and celebrations within the Left and among the young, the rest of society remained more or less status quo. As one character in a farcical Theater of All Possibilities play put it, "they stuck a sugar cube in my pinkies and said, 'Turn on and tune in.' Then they said, 'The decade's over, man, the Beatles were a millionaires shuck, the others O.D.'d, the Big Man himself's a stand up comic, and Yip-Yap's on Wall Street.'"[20]

As a result, many young idealists gave up on changing the whole world and headed for the margins of society to start their own little worlds. At Synergia Ranch, they found a place with order and meaning—a strange new order, perhaps, but order nonetheless. There they could finally live out their values, along with like-minded people. "The search for truth was really important. And truth between the people," said Phil Hawes. "It was really important to be looking for the truth, whatever that might be—which is hard to find, but maybe that was part of the idea."

The ranch members were working with dirt bricks and strong motivations; they believed their daily labor was connected to the planet's future. To create a "synergetic civilization," the ranch members agreed, they must transform themselves. They threw themselves into a full life of work, study, and theater. For their minibusinesses, the residents started a variety of economic enterprises; everyone had to support themselves. Some pooled funds to build a pottery shop where they made ceramic dishes and tiles. Some made jewelry. Others started a printing press. In a woodshop, residents carved sturdy wooden tables and chairs, windows, and doors; they sold some to people passing through, and used the rest to furnish the ranch. "We built our tables in the old days in New Mexico really strong so people could dance on them, or we'd occasionally have fistfights on them when people were having problems with each other," recalled Phil Hawes.

All that physical activity was just one surface of the ranch's life, however. Much of the real work was happening metaphysically. Twice a week the residents gathered for meditation sessions. Nightly dinners also became an important part of ranch society. Most nights, everyone ate together in silence, until at the end of the meal, John or another moderator called out for "comments and observations" from those seated in the dining hall. After Sunday dinner, each person made a short speech about his or her week. And at two special dinners every week, the residents educated themselves. Tuesday night dinner usually featured literature or culture from some part of the world. Everyone read an assigned text beforehand, then came to dinner in costume as a character from the reading; cooks prepared special foods to match the night's theme. On Thursday nights, John Allen gave after-dinner metaphysical talks.

A broad spectrum of subjects came up at the sturdy wooden dinner table, and everyone at the ranch was expected to master an eclectic informal canon of texts and ideas deemed necessary to the synergetic civilization. "It was a very structured existence," Phil Hawes recalled, "but we learned a lot of good habits—how to look at things, I think, to see if they're really real, or if it's total bullshit." The synergians read Idries Shah's Sufi tales for bits of wisdom. They studied Lewis Mumford's *Technics and Civilization*, which elaborated the history of technology and laid out a blueprint for small, self-sustaining communities of the future. They discussed W. R. Bion's *Experiences in Groups*, a treatise on group psychology, to analyze their own interpersonal dynamics. For example, according to Bion's typology, were they behaving as a "fight or flight" group, avoiding the task at hand by becoming absorbed in conflict? Or had they become a "kill the leader" group, derailing themselves by attacking the person in charge?

The goal of all this study was not just to accumulate knowledge; it was to achieve personal and collective transformation. To birth a new world, the ranch residents believed, they had to become new people. "We were trying to become artists, scientists, and adventurers," said Phil Hawes. That combination was not random. By developing themselves in those three ways, John Allen taught, the synergians would get in touch with their three "centers": emotion, thought, and action. Then, and only then, would they be fully developed and balanced people, capable of awakening. Like most of the ideas discussed around the dinner table, that idea was not new, but salvaged from the old civilization—in this case, from an early twentieth-century mystical guru named George Ivanovich Gurdjieff.

> As he grew up, his urge to understand the meaning of human life became so strong that he attracted a group of "remarkable men" . . . From the time of his return to the West, he worked unceasingly to gather round him a group of people ready to share with him a life wholly turned towards the development of consciousness. He unfolded his ideas to them, sustained and gave life to their search, and brought them to the conviction that, to be complete, their experience must include at one and the same time all the aspects of a human being.[21]

That passage could have been used to describe John Allen. In fact, it was a description of Gurdjieff, from a volume of his stories. There were striking parallels between the two leaders' lives. Born in Russia around the 1860s, Gurdjieff journeyed through the Middle East and Central Asia, apprenticing himself to holy men of all traditions. When he returned to Europe, in the 1920s he founded his Institute for the Harmonious Development of Man in the French countryside, and gained a cult following as a spiritual teacher. Residents followed demanding schedules of work, dance, and personal development work with the guru. In a voluminous trilogy he modestly titled *All and Everything*, Gurdjieff expounded his theory that most human beings are not fully awake, living as if they were machines. Few people fully develop their capacities for feeling, thought, and action, Gurdjieff argued; most people are underdeveloped or lopsided. The highest goal of life, he taught, was to develop all three "centers" in order to become a more fully evolved human being.

Gurdjieff's ideas rippled outward through the European counterculture and to America. His disciples called their practice simply "The Work." John Allen never considered himself a Gurdjieffian, "because the Gurdjieff thing is a very orthodox thing; and John, one thing he's not, he's not orthodox," as Ben Epperson put it. But Allen and some of his friends did meet John G. Bennett, one of Gurdjieff's British disciples, and were swayed by some of the ideas. At Synergia, the residents carefully read John Bennett and other Gurdjieffians' work. Bennett had written in 1949: "We are passing through a period of transition in human history when, in some sense, an old world is dying and when, therefore a new world must be born." Further, he argued, a few awakened people must, through intense thought and effort, help that new world evolve into being.[22] That dictum resonated deeply with many of the young Synergia residents.

However, life at Synergia was not all devoted to work and study. It was equally about play—specifically, theater. On stage and in theater practices, the ranch residents were learning to toy with their identities and relationships, to work through historical patterns, and to experiment with future possibilities. Indeed, that became the name of their troupe: the Theater of All Possibilities. Every Saturday and Sunday morning, ranch residents gathered for theater practices, engaging in group exercises and rehearsing plays.

Decades later, as they tried to explain to me the value of theater, each actor had a slightly different interpretation. But to all of them, acting was an essential part of life—not just a hobby, but a way of being and becoming all that they wanted to be. John Allen said theater "kept people from becoming ideological" as people playacted out their own philosophies. Allen himself was not above such self-mockery, writing original plays such as "The Guru" and "McNeckel's Commune," which poked fun at communes of the time. Theater also gave the actors a safe space to play with the intense interpersonal energies that arose from living together. "Theater gives you a certain distance to make fun of it," recalled Mark Nelson. "You can express all your group frictions."

Theater also formed a central part of the ranch's human development laboratory. "It's the development of the emotional center," said Bernd Zabel, using Synergia-speak. "It's a very fascinating way to live,

I mean, to experience totally new emotions." Theater "trains your emotional center, and that teaches you how to move, and how to talk; you can work on all sorts of interpersonal interactions during theater exercises," said Ben Epperson. During our interview, he suddenly jumped up in front of me to demonstrate theater exercises. An old man, he still remembered all the movements, swaying his lanky frame, swinging his hips around, then screwing up his face in all sorts of contortions, while bouncing his arms out suspended at his sides. Exercises like this were more than physical, he explained, speaking rapidly and excitedly: "At the same time you're moving and you're using your mind, your intellectual mind to check out your movements, and you're exploring the small muscles of your face . . . The idea is to balance the thinking, feeling, sensing."

Daily life itself became its own theater as ranch residents took on alternate identities. Some went by nicknames instead of their given names. John Allen was "Johnny" (later becoming "Johnny Dolphin," his pen name). His wife, Marie, was "Flash." Their longtime friend Kathelin, the Theater of All Possibilities director, was "Honey." This tradition of renaming, claiming new identities, would continue up to the days of Biosphere 2.

The actors sold tickets in the streets of Santa Fe, and the Theater of All Possibilities made its grand debut. The actors put on their first performance on a snowy winter day in 1970, outdoors in the ranch courtyard; they served the audience bottled brandy to keep warm. In their wild first festival, they performed seven different plays in seven days. The troupe would go on to put on plays by every great theater tradition, jumping from the ancient Greeks to Shakespeare to Goethe and Molière to Bertolt Brecht; from Japanese Kabuki to Hindu kathakali dance-theater to colorful European expressionism. There was a purpose to the eclecticism: they were "collectively getting a sense of what's been done before," explained Kathelin, the director. The theater company's ambition, declared its published collection of scripts, was "to explore man's intentions toward the planet and the galaxy, as Shakespeare explored man's intentions toward England, Goethe toward Europe, and Brecht toward economic history."[23]

Thus, to the leaders of the theater troupe, the ultimate purpose of acting went far beyond the actors' emotional development. Theater, to them, was about nothing less than working out their place in history. "The aim of Theater of All Possibilities was always to investigate and to perform pieces that have a historical context applicable to the present," recalled Kathelin. "What we call history has come about through a series of decisions, which are sometimes very private and yield larger consequences. This makes the study of theater useful for people who are interested in conducting enterprises that aim to have a historical relevance and social implications." On stage, the actors could play with the past and the future—acting out what had already happened in human history, and preparing themselves to act in real life. By staging both classic works and also original plays by "Honey" (Kathelin) and "Johnny" (John Allen), the actors were playing with the idea of synergetic civilization: to bring together the best of the past, and out of it create a new future.

The actors began putting on festivals at the ranch every summer. They built their own theater, a high-ceilinged geodesic dome in the style of Buckminster Fuller. For their traveling headquarters, they bought a secondhand yellow school bus, strapped their props and costumes to the top, and began touring the country off and on, first as the Theater of All Possibilities, later under the name Caravan of Dreams. Money was tight; the actors paid their own way and equally split whatever money they got at the door. Their tours started humbly. For the first audience, "We went out to a little town in Nevada, and I think we had three people, two of whom were honorary ticket people we got from the streets," said Ben Epperson.

The theater tours began in the United States but progressively traveled farther. In 1971, the troupe toured Western Europe. On other trips they visited Canada, Mexico, Nigeria, Persia. They chose plays each spring, rehearsed in the summer, and toured in the fall, which gave ranch residents some time to arrange their minibusinesses and finances in order to get away. At the end of each tour, a few actors would keep on traveling to check out the greatest hits of past civilizations on the way: through archaeological sites in Greece, across the Sahara, around Iran, across the Khyber Pass, through mining communities in the United States. "The venues of our tour performances were as various as you can imagine," recalled Kathelin. "From a theater on the West End in London to outdoors on the volcano Popocatépetl in Mexico . . . in squats, in academies of art, for scientists on sand dunes at White Sands, in a Sufi *tekkia* in Istanbul, in temples in Bali, for Aboriginal communities in the bush, a voodoo church in Paris, university stages, warehouses, street theatre for the homeless, music clubs, at Second City in Chicago—the list goes on and on."

In still another sense, when I took a step back from the particulars of their story, it seemed the synergians were acting out a broader drama. In spite of their feeling of doing something new, they were reenacting a sort of unwritten script, a deeply ingrained cultural pattern. The drama had played out repeatedly in European and American history: the quest to abandon a problematic "old world" and to create a new world in line with higher ideals. Even though they felt themselves to be rebelling against mainstream American culture, in fact, by heading out to the frontier to create a better life of their own, the synergians were reliving the very creation myth of America itself.

In fact, Synergia Ranch—and Biosphere 2—thus had historical roots much deeper than the late twentieth century. Interestingly, John Allen and his closest associates reacted explosively when I told them I was going to include a discussion of utopianism in this book. "Say that John Allen believes Utopias are impossible and that it's ridiculous to use an idea of Utopia in design, that he considers Communes a snare and a delusion," John wrote to me. "The history of Communes show[s] how limited (in time and space) the Utopian approach is. Your appraisal of them is correct; they can only survive as local authoritarian communities. That's why I had and have nothing to do with them; they don't work well or long. I believe individual and corporate enterprise are the highest forms of human organization."[24]

Yet when I looked at the magnificent Biosphere 2 structure, its rainforest pyramid glimmering in the sun, I could not help but think of Western culture's perennial dream of an ideal new world. Likewise, when I listened to the reminiscences of those who had lived at Synergia Ranch, many struck chords that resonated with the new world aspirations of American communes. Sociologist Rosabeth Moss Kanter, in a comprehensive study of American communes, found that almost all communes shared a general two-part orientation toward the world: (1) the outside world was wrong, unjust, unnatural; but (2) together, in a minisociety, the commune-dwellers could make it right. In that vein,

Synergia resident Randall Gibson recalled, everyone at the ranch shared a sense that "society is so strange right now"—when it seemed that anyone of any revolutionary consequence in the United States was getting shot—so "we had signed ourselves over to a new foreign legion, and John was our post commandant."

Most crucially, Kanter's study found, it has always been important for commune participants "to believe that life is an expression of their ideas, that there is no separation between their values and their way of life."[25] That belief resounded over and over in the stories of Synergia Ranch residents. As Kathy Dyhr put it, "I liked the fact that there were intelligent people, and that ideas were the primary thing guiding what everyone was doing . . . to look at different ideas, but to actively design a lifestyle based on a philosophy and based on an idea or on a series of ideas . . . making your life a work of art."

Why then did John Allen and his friends so strongly resist labels such as "commune" and "utopia"? Part of the reason was undoubtedly the reaction of people tired of having their science project bashed in the press simply because the team who built it had lived an unusual life together. But perhaps more significantly, by the late twentieth century those had become dirty, belittling words. Anyone who believed in the possibility of a perfect world must be ridiculous. Yet visions of a longed-for ideal world reach as far back as the beginnings of Western civilization, to the Judeo-Christian belief in the Garden of Eden (and an accompanying hope for paradise's return). Though cultures all over the world have imagined paradises, Western civilization has birthed the most myths and theories about the possibility of actually creating a heaven on Earth. As Frank E. and Fritzie P. Manuel wrote in their definitive history of utopianism, "There are treatises on ideal states and stories about imaginary heavens of delight among the Chinese, the Japanese, the Hindus, and the Arabs, but the profusion of Western utopias has not been equaled in any other culture."[26]

Indeed the history of the very term *utopia* reflects the slipperiness of the concept. In 1516, Thomas More coined the word as the name of a fictitious island society where everyone lived in perfect harmony. "Utopia" was More's play on words. It could be traced to Greek roots to mean either "the good place" (*eutopia*) or "no place" (*outopia*)—implying that the fantasy was a good one, but perhaps an unrealizable one as well.[27]

Nonetheless, following More's lead, European thinkers and writers produced a whole genre of utopian literature. They wrote detailed descriptions of imaginary perfect societies in faraway lands, from Margaret Cavendish's woman-centered society in *The Blazing World* to Sir Francis Bacon's *New Atlantis* governed by scientists. Utopianism became a major chord in secular European philosophy as well. Everyone from Enlightenment theorists to Karl Marx wrote out theories of how to enact the perfect world, whether through reason or through revolution; they all seemed to assume it was actually possible.

The very beginnings of European settlement turned North America into a stage for visionaries trying to act out their imagined ideal worlds. Offering plenty of land and comparative religious and economic freedom, the continent seemed a blank slate for new experimental minisocieties—literally the "New World" so many had waited for. Would-be prophets and outcasts jumped on boats bound for the new continent, determined to set up miniature societies in line with their values. By the 1970s, American communes would be seen as outsiders or part of a "counterculture"—but many of the first white settlements on the continent were in fact utopian communes. The fabled Pilgrims, who established Plymouth Colony in 1620, worked and held property communally, as did other Puritan settlers. They and other religious dissident groups sought to form self-sustaining colonies where they could live a pure religious life. Variations on those early communal arrangements would be repeated over and over in American history. As historian Donald E. Pitzer has argued, "the idealism of these early formative experiments has continued to influence the development of America's distinctive sense of national mission."[28]

As the American frontier moved west, so did the communal utopian experiments, from Midwestern religious communes in the nineteenth century to West Coast hippie communes of the twentieth. The desert attracted some extreme dreamers, searching for open space in which to live by their own rules; one of the most blatant examples was the Mormons' colonization of Utah. In the post-sixties commune movement, another ambitious utopian minisociety that germinated in the Southwest was Arcosanti. Founded in 1970 in the desert north of Phoenix, Arizona, the cement-domed minicity was the brainchild of visionary Italian architect Paolo Soleri. Kindred spirits to the synergians with their ideas of "ecotechnics," Soleri and his followers believed that through "arcology"—blending architecture and ecology—people could create a new basis for cities, and thereby create a new basis for society.

But even if humans could envision an ideal world, could they really enact it? Synergia Ranch was far from a typical hippie commune. Yet some key motifs in communal history can help to explain Synergia Ranch—and in turn, the tale of Synergia Ranch illuminates several key lessons in the history of utopian communities. The New Mexican ranch illustrated one of the most important facets of successful communities: a shared personal belief system. In many of the longest-lasting nineteenth-century American communes, shared religious beliefs lay at the core. In one of the foremost success stories, the Shakers in 1794 followed their leader "Mother Ann" Lee from England to America, because she had a vision that the place for a new religious world had been readied in America. Viewing the outside world as diseased with sin, the Shakers believed that by living together, separate from the outside world, confessing their sins, and remaining celibate, "they could become perfect."[29] That vision appealed to so many that by 1850, the sect had grown from nine members to four thousand people living in eighteen settlements from Kentucky to Maine—even in spite of the celibacy requirement that necessitated a constant stream of new converts. Other religious sects such as the Amana and the Hutterites also thrived in communes, believing that they were living a religiously pure life, in contrast to a disorderly outside world.

Obviously, Synergia was not a religious sect like these in any way, shape, or form. But not all successful belief-centered American communes have been religious. For example, socialist commune founders had a similarly passionate relationship to ideas; socialist philosophy was their new god. Socialists founded their share of utopian living experiments in the mid-1800s. Some of these were short-lived, as book-smart socialists often proved lousy farmers. But many secular communes, like the later Synergia Ranch, thrived on an intensely spirited devotion to shared ideals. Furthermore, the socialists followed the same social pattern as the religious communes in the way they clung to prophets; the longest-lived communes, whether religious or political, usually followed some visionary leader. Followers of French socialist Charles Fourier founded twenty-eight Fourierist "phalanxes" in America during the 1840s and 1850s, with a flagship commune at Brook Farm, Massachusetts. All of the phalanxes followed Fourier's elaborate system of communal work and living. In Indiana, followers of Scotsman Robert Owen strove to create a "New Moral World," as Owen called it. Under Owen's direction, they established an agricultural and manufacturing village called New Harmony, based on socialist principles and a system of lifelong education. The children there sang of a return to an almost biblical-sounding paradise:

> The race of man shall wisdom learn,
> And error cease to reign:
> The charms of innocence return,
> And all be new again.[30]

Even in that secular socialist community, an almost religious sense of meaning derived from members' belief in the possibility of a "new" world.

Still, that sense of shared belief, evident at Synergia and in many contemporary communes, is not usually enough on its own to sustain a community. Rosabeth Kanter, in her study of "commitment mechanisms" in intentional communities, found that most successful communes share other structural traits. Members are bound together not just by shared beliefs but by strong social structures: the requirement of intense commitment and sacrifice from each person, strong leadership, set rituals, and highly structured social and economic relationships among the people. All of these factors were present at Synergia Ranch. Another researcher, Benjamin Zablocki, studying communes over time, found a correlation between a community's longevity and the presence of strong leadership.[31] Moreover, Zablocki noted that in flourishing communes, authoritarianism tended to increase over time. Ironically, many successful communities, though they were supposed to be about cooperation, depended on a strong leader. The same was true, of course, at Synergia Ranch. The charismatic John Allen, in his forties, was older than most Synergia residents, in their twenties. Many key people worked to set up and run the ranch, and John was not always present, as he often traveled for other ventures. But though the early atmosphere of the ranch was organic and spontaneous, "Johnny" was still the main leader: teacher, moderator, lecturer, author of scripts both onstage and off.

Communal history—including the history of Synergia Ranch—suggests, then, that people function together best in a highly structured environment. As much as they need meaning, they also need practical instructions about how to live. In contrast, most 1960s communes offered no real alternative structure to mainstream society; seekers knew what they were running *from* but had nothing specific to run *to*. With its full life of planned activities, Synergia Ranch was, in a sense, a nineteenth-century community living in the 1970s. Many New Age communes, which arose out of a critique of authoritarian society, eventually dissolved, lacking any authority structures at all. As one disappointed contributor to *Modern Utopian* magazine wrote, after relearning the lessons of history for himself,

> Dear Friends,
>
> I have spent over twenty years in what is now called the "Hippie Movement," living in short-lived communes based on anarchistic freedom and long-lived communes based on religion . . .
>
> First, those communities based on freedom inevitably fail, usually within a year.
>
> Second, those communities based on authority, particularly religious authority, often endure and survive even against vigorous opposition from the outside world . . .
>
> If the intentional community hopes to survive, it must be authoritarian, and if it is authoritarian, it offers no more freedom than conventional society. I am not pleased with this conclusion, but it now seems to me that the only way to be free is to be alone.[32]

This was the sad paradox many encountered in efforts to found an alternative society: If commune-dwellers tried too hard to resist conventional models of leadership and authority, they often fell apart for lack of direction as members drifted away. If they wanted to hang together, however, they ended up imitating the hierarchical social structures they had grown up with. Nowhere was this more clear than at Synergia Ranch. As the anthropologist Laurence Veysey observed, this "group in the desert, despite its great pride in creating an intricate formula for balanced living, was grotesquely dominated by the tough, achievement-oriented values of mainstream America."[33]

The enigmatic John Allen inspired many young people. It wasn't just that he could speak for hours about philosophy, science, history, and a host of world cultures; he could convince his listeners that their own shared lives mattered to the future of the whole world. Even more than his ideas, Allen's energy drew people to him. Decades later, at Synergia Ranch, I still noticed it: when he walked into the room, even if he was quiet, somehow it was impossible not to be aware of his presence.

Many considered John Allen to be brilliant, but he was also volatile. Like a deeply focused actor, at any moment he could burst out with hearty praise, humor, ponderous musing, or piercing anger. "He was one of those charismatic figures who could be disarmingly charming, or, as I came to learn, cut you down with a glance," later wrote a biospherian, Jane Poynter. "His smooth face could explode with the enthusiasm of a kid at Christmas, or tighten into a sardonic glare that would have made Mephistopheles proud."[34] At after-dinner discussions, John might suddenly and loudly attack anyone at the table. Seemingly unprovoked, he would pick out some ranch member to publicly berate and humiliate for any faults, while the rest of the synergians sat quietly by. Numerous residents recounted such experiences; some seemed obsessed with John's two-sidedness even decades later. Randall Gibson was one of those. Even two decades after leaving Synergia, he spoke of his former mentor with equal parts awe and anger: "John Allen has a brain muscle which is large as the Belgian Congo, and as dark, and as conflicted." One of Synergia's earliest members, Gibson left the ranch alienated and angry. He became convinced that theater was less a personal development tool than a "very powerful set of emotional technics, a part of the behavioral control laboratory of this group."

No one was being held at Synergia Ranch by force; the largely intelligent and motivated group of people simply tolerated all sides of John Allen. In the Gurdjieffian school of thought, which was one strand of the Synergia curriculum, a teacher's negative feedback was seen as part of each person's growth process and therefore not to be questioned. In that light, ranch members generally accepted John Allen's harsh outbursts.

In fact, Randall Gibson recalled, Allen's loudest tirades—and, from time to time, even physical blows—were seen as a special challenge for the person on the receiving end, designed to "make them hit a higher breakthrough." It was just part of life that "every now and then you'd have to be officially excoriated," Gibson said.

Yet despite Allen's central role, a spirit of cooperation, and a sense that each person was important to the ranch's work, did shine through, Gibson said. "Everybody there was there to take a risk and do something new . . . It was 'rah rah, we're all doing this together.'" Phil Hawes agreed: "It used to be that I would say to John or to different people, 'You know, if I could find anything more fun, I'd do that, but this is the most fun I've ever had in my life.' With the hard parts, the unpleasant parts, learning about yourself . . . you get used to that; you sort of like that."

Somehow Synergia's idealism and enthusiasm, from the beginning, coexisted with the seeds of conflict; cooperation coexisted with control. In that the synergians were not alone; they were merely following patterns long established in the history of attempts to build a "new world." These would be the seeds of Biosphere 2 as well.

And so the ranch residents began to build. Several of them entered a joint venture to form a construction company, which they called Synopco, short for Synergetic Operations Company. For their first major project, in 1974 they constructed an adobe house in Santa Fe, using the methods they had tested on the ranch. They formed earthen walls and carved the house's wooden doors by hand. The architect was Phil Hawes, future designer of Biosphere 2. In more ways than that, the adobe house in Santa Fe was a preparation for the future. To Stephen Storm, who served as foreman constructing the house, completing that first building project meant more than just a job. It felt like proof that together, through hard work, he and his friends really could accomplish anything they wanted. He recalled years later,

> Probably the most important events of that first period were when we formed a construction company and built our first house . . . It was the experience of building that house, that really I guess helped me to see the potential that a group of people working together could accomplish. Because we really

> didn't have anything—we had no money, or very damn little, and basically we were able to go into Santa Fe, and walk away from a completed house. And I think that was really important for everybody, because that was the beginning of the line of work that has wound us up here [at Biosphere 2], able to build an enormous structure. Really the seed was that house on Camino Manzano which we built then.[35]

Thus the synergians began to believe that they could make all possibilities real.

CREATE AND RUN

A CEMENT-HULLED SAILING SHIP CRUISING THE WORLD'S OCEANS. AN organic farmhouse estate in southern France. An enormous ranch station, isolated in the Australian Outback. A brick hotel in Kathmandu. An avant-garde art gallery in London. A rainforest restoration effort in Puerto Rico. A multilevel performing arts center in Texas crowned by a geodesic desert dome. Over the next two decades, these would become the synergians' frontiers. On these stages, a growing group of aspiring scientist-adventurer-artists would try to redefine the human place in nature. Equally important, they would learn how to act together.

As had been the case from the beginning, the impulse to expand came from John Allen's itchy restlessness, surfacing even in the midst of success. In 1974, Synergia Ranch's construction company, Synopco, was carving out a niche for itself in Santa Fe. Using their earth-building skills, the ranch members broke ground on a new commercial project, an all-adobe thirty-unit condominium complex in town. By rediscovering traditional adobe building methods, the crew were actually ahead of their time; Santa Fe's characteristic architecture superficially mimicked the indigenous earthen-walled dwellings of the Southwest, but the synergians were some of the first local builders to actually return to using the original natural materials. As they worked together, they were building more than clay houses—all of the eventual board members of the company that would build Biosphere 2 had some role in the adobe condominium project, as builders, designers, or executives.

But, like so many times before, John Allen's mind was already traveling further. Marie Harding was shocked when her husband bluntly announced to her one day that it was time for them all to take off and build a ship to voyage the oceans. After the theater group's multiple moves cross-country, ranch life had just begun to take on a stable rhythm in New Mexico. But John was convinced. After all, the planet was mostly ocean. Therefore, he reasoned logically, if they were really going to become part of a global "synergetic civilization," they needed to become "sea people" as well:

> Two-thirds of the atmospheric surface of the biosphere touches ocean, and without the water cycle driven by sun and ocean, the rest of Earth would be desert . . . I knew that without a world-ranging ship that would expose me and the creative group to the vicissitudes and glories of the ocean, my talk about understanding biospherics and eventually building Biosphere 2 would become fakery. Just endless chatter . . . I had to become a flexible being that directly experienced Planet Water.[1]

And so the synergists' planetary adventures began. Several ranch members piled into the theater company's broken-down yellow school bus, with a fifty-five-gallon drum of oil strapped to the top to keep lubricating the oil-guzzling engine. They headed west, toward the sea. When they got to Berkeley, California, they moved in with the mother of one of their members, the quietly brilliant physicist and mathematician William Dempster (future engineer of Biosphere 2). Later on, the builder/actor group would move down into a tin shack by the docks in Oakland. There, at the bay's edge, they gathered materials and built their ship. For seven months they labored intensely, seven days a week, taking half of Sunday off. At one point they almost gave up and drove the bus back to New Mexico. But as with their adobe projects, the most important building resources were determination and many hands. Together fourteen people formed the boat's hull out of ferro-cement, first sculpting a shell of wire mesh and then coating it with a thin layer of cement—a construction style requiring lots of labor (which they had) and little money (of which they had less). Half of the building crew would later be involved in building Biosphere 2.

They modeled their vessel after the style of a Chinese junk, a classic old sailing ship that John Allen had encountered on his Asian travels. The entire project cost $90,000, including $60,000 that group member Randall Gibson, heir to a family fortune, contributed out of his own trust fund. The builders worked twelve-hour days without pay in exchange for "sweat equity" ownership—those who sweated together would own the boat. To support themselves, they opened a café in downtown Berkeley, The Junkman's Palace, serving hamburgers to the resident bohemians.

The head of the construction crew was only twenty-one years old. The sharp-tongued, fair-skinned Margret Augustine was a new member from Canada. She supervised her fellow shipbuilders and sewed the sails by hand. She had dropped out of high school, and, like the rest of the building crew, she had never built a boat before. But "Firefly," as the quick-tempered Margret became known in the group, had a strong will, and she had an instinct for organizing people. One day she would become CEO of Biosphere 2.

"She had real managerial ability . . . At her best, she was a very good people leader, and dynamic, and everything," recalled Ben Epperson. "At her worst, she could be extremely dogmatic and defensive, aggressively defensive, defensively aggressive." Like many of her fellow builders, Margret had followed a circuitous route into this crew. She had been planning to travel to Asia on a personal quest; she and her boyfriend posted a sign in a Berkeley bookshop saying, "If you are interested in working on yourself and are willing to make sacrifices, contact us," Ben Epperson remembered. So, he said, "I went and I talked to them. They were planning to go East, and I said, 'Well, that's great. But I know an American group that works hard, and has some intention . . . Before you go to India or wherever you're going, why don't you drop by Synergia Ranch?'" Margret Augustine was hooked.

At last, in the spring of 1975, the ship was ready: eighty-four feet long, a shining black-painted hull holding cabins for up to fifteen people, three towering masts billowing the hand-sewn canvas sails. The builders dubbed her the Research Vessel *Heraclitus*, after the ancient Greek philosopher who proclaimed the ever-changing nature of the universe: "You can't step into the same river twice." In that spirit, the RV *Heraclitus* set out into San Francisco Bay, sailing under the spires of the Golden Gate Bridge and out into the Pacific Ocean. Randall Gibson remembered the crew's glee on the deck: after "the synthesis of a lot of effort," at last, "a whole bunch of crazies go sailing out into an ocean in a ship they built together!" They decided to sail west to Hawai'i, then onward to Southeast Asia and the Indian Ocean. But on their first night on the sea they met a terrifying storm, so they shrugged and made a group decision to go around the world the other way, heading south, then through the Panama Canal, and eventually across the Atlantic to Europe.

Meanwhile, some of the most important future-determining events in the group's life were happening more quietly. The same year that the ship was under construction, another young, mild-mannered idealist arrived at Synergia Ranch. His arrival would prove more significant than anyone there could have imagined upon first meeting him. He had dropped out of architecture graduate school at Yale and moved to New Mexico in 1970. Like so many of the other synergians, he yearned to really do something with his ideas. "I'd spent, what does it come out to, eighteen years in school. And to get out and practice things and to learn things by doing them and so forth, at that time was tremendously stimulating," he said.[2] His name was Ed Bass, and he and his three brothers happened to be heirs to one of the richest oil fortunes in Texas. Through shrewd investment, the brothers were well on their way to becoming one of the wealthiest families in the country. While capitalist older brother Sid managed the family finances, Ed, the second brother, wore a ponytail and told friends he craved a simpler, fulfilling life. Ed had grown up with a love of nature. As a boy, he had enjoyed fishing trips with his father, weekends riding on the family's cattle ranches, and vacations on a small coastal island owned by his uncle, oil tycoon Sid Richardson.

After dropping out of graduate school at Yale, Ed set out to start fresh in New Mexico and began building his own house in the mountains outside Albuquerque. But the project stayed half-finished, out of compliance with building codes, and the state was threatening to tear it down. By chance Ed attended a theater performance at Synergia Ranch. Around the same time, he also got to know Phil Hawes, the Synergia earth architect, through a mutual friend. Soon after, Ed hired the ranch's construction crew to finish his house. The relationship was further cemented in 1974, when he entered his own construction company, Badlands Conspiracy Co., into a joint venture with Synopco, to finance the adobe condominium complex in Santa Fe. Shortly after, Ed finally moved to Synergia Ranch. When not working on building projects, he learned to act and toured with the theater company. Though naturally shy, he enjoyed being onstage; on the Theater of All Possibilities African tour, he played a farcical capitalist character who ate bureaucratic red tape, and the Nigerians loved him. It was the quiet beginning of a business relationship that would grow to span two decades and hundreds of millions of dollars.

The resources had come. Now came time to "create and run." John Allen had grabbed that motto, like many other bits of wisdom, from the Sufis. Keep building, keep energy high, moving on and on to higher challenges, scarcely stopping to look back—this, Allen believed, was the road to greatness. He would later describe his management style: "Nothing fails like success wise men say, because from success a man will often forget how to make the next step, and but plunge down from a higher precipice than before; therefore, the wisest men say, best to succeed, to go from achievement to deeper achievement until death confronts each of us."[3]

"We would cram so much into a period of time that there was a strange dislocation of time-sense," Phil Hawes recalled. Each year new projects began. More members joined Synergia and kept the ranch humming while others fanned out into the world. The Theater of All Possibilities hit the road for new tours every year. The ship coasted around the world, as young crew members hopped on and off at ports all over the planet. The boat's voyages would eventually include a three-year round-the-world journey through the tropics; an ethnobotanical mission two thousand miles up the Amazon River to collect plants with shamans; and an expedition to Antarctica.

Meanwhile, as the *Heraclitus* sailed the seas, friends back on the ranch were building their biggest conceptual vehicle for creating and running: an organization they called the Institute of Ecotechnics (IE). Mark Nelson, the small, smart Dartmouth philosophy graduate with a gift for eloquent speeches, became the Institute's first chairman in 1973. (Except for 1975–82, when Randall Gibson was chair, Mark would be the IE chairman for three decades to come.) At first the Institute of Ecotechnics was little more than a name on paper, in which various members awarded themselves official titles. But the group's huge aspirations were reflected in the very name "Ecotechnics," a word of their own invention. The name bespoke a dream: that they could harmonize ecology and human technology. At the time, many 1970s environmentalists were battling against the negative aspects of human technics. Contemporary American environmental legislation focused on restricting pollution of the nation's air and waters. Doomsday reports such as the influential *Limits to Growth,* published by the Club of Rome in 1972, proclaimed the alarming dangers of an expanding global human population. Humans seemed to be on a collision course with the natural world. But the new self-styled ecotechnicians were convinced they could find ways to save wild nature and embrace human technology all at once. One of John Allen's intellectual heroes, Lewis Mumford, had written in 1934 that the next stage in history would be a "biotechnic period, already visible over the edge of the horizon"; this biotechnic society would use life forms, rather than artificial machines, as the basis for survival and invention.[4] Mumford envisioned a world of small, self-sufficient mini-societies, each living in ecological balance within its own region. The members of the Institute of Ecotechnics took his message to heart. They believed they were expanding Mumford's concept. As Mark Nelson argued, "the reconciliation between technics needs to be not only at the individual life form level (bio) but at the ecological or ecosystem level (eco)."[5] The participants also believed they could create a revolutionary way to live. As three of the Institute's directors later wrote, they aspired to create "a new discipline . . . critical for themselves personally, their society, and for Nature."[6]

"Create and run" was exactly what the Institute of Ecotechnics did. The title of a chapter in John Allen's memoirs—"Ventures in Timber and Jazz"—hinted at the wildly broad scope of the various projects that the group would soon take on. In the mid-1970s, with young Ed Bass's millions close at hand, the Institute's actors-cum-builders suddenly had the freedom to turn theories into action. John Allen, the businessman, set up a corporate structure: Bass as primary investor, the other key group members as managers who would hold "sweat equity." The Institute's board members included Marie Harding, Mark Nelson, John Allen, Ed Bass, and theater director Kathelin Gray—all of whom would eventually sit on the board of Biosphere 2. Though everyone on the IE board lived closely together, the new arrangement on paper was "very, very businesslike," recalled Phil Hawes; they went over their business agreements every year.

Now they would begin to assemble a world of their own, piece by piece. The Ecotechnic team located and bought new properties in rapid succession. As biospherian Jane Poynter put it, "when the members of IE got together and decided to do something, they didn't sit on their rear ends and think about it interminably."[7] Within a few years, they were working and acting on four continents. First, the directors decided, they needed a home base for their European theater tours. Drawn to the creative artistic history of southern France, a small team scouted the countryside in a bus in 1975. They put on business clothes for real estate meetings during the day and slept in the bus at night. They found a beautiful eighteenth-century stone farmhouse outside Aix-en-Provence and decided it was perfect; Ed Bass threw in half a million dollars, and the Institute bought the house, naming it Les Marronniers, the French name for the picturesque old horse chestnut trees found there. More Institute of Ecotechnics/Theater of All Possibilities members arrived; new ones came along and joined. They remodeled the old mansion to create art workshops and performance rooms. They started an organic farm and orchard on the grounds. Most important, they made the house and gardens into a gathering space, where eventually they would hold conferences.

Next stop, South Asia. John and his wife, Marie, fondly remembered their explorations there. At the foot of the Himalayas, Kathmandu had become an international crossroads, magnetizing young travelers and enlightenment-seekers from all over the world. There, in 1977, the ecotechnicians got together with a group of Tibetan refugees to begin their new construction project: the red brick Hotel Vajra. But it was not to be just a hotel. They envisioned it as a fusion of East and West—a site in the "synergetic civilization." The Vajra must be a cultural crossroads, John Allen wrote, "a world which would serve as a gateway into Mountain Asia, The Rooftop of the World, a gateway that people could enter from all directions and go out in all directions."[8] There, scholars, explorers, spiritual masters, and other creative people from all over the world would meet on their travels, relax, and explore each other's ideas. Again Ed Bass put up half a million dollars in seed money to buy land and materials. Phil Hawes sketched the plans, the group of Tibetan friends helped find land and workers, and Margret Augustine—"Firefly"—again swung into action to manage the multicultural crews of builders and artisans. A master Tibetan muralist painted the pagoda ceilings, and Nepali craftsmen carved intricate dark wooden doors and ornamentation. The builders wrote to the Dalai Lama to request the books of the Tibetan Buddhist canon for the hotel's library and received Hindu canonical texts from a local swami. For the Hotel Vajra took its name from the Sanskrit word for "thunderbolt," the Tibetan Buddhist symbol of the power of the enlightened mind.

In the late 1970s, the growing Ecotechnics crew was doing more than just acquiring properties around the planet. They were playing with possibilities, testing their ability to realize their visions. They were also learning to work with a world of different ecological and cultural situations—gathering the pieces they would need for their own new world. In the spirit of "create and run," back in New Mexico, while Synopco was still finishing off its four-year adobe condominium project, some of the construction crew began taking horse pack trips into the mountains around the ranch after hours. They were training themselves for their next mission. Then, in 1977, a delegation took off to Australia's Outback frontier. They took on (again thanks to Ed Bass's massive bank account) an enormous tract of land, Quanbun Downs, 300,000 acres of open pastureland in the sparsely populated Kimberley region of northwestern Australia. Land was cheap there for a reason: in the hot, tropical Kimberley, as the local joke ran, there were three seasons: flood, drought, and fire. Past human mismanagement had made the land even harsher, as overgrazing ravaged the native grasslands. Charging forward, the posse of synergians who moved there turned themselves into ecocowboys (and ecocowgirls). They determined to learn to run cattle on the savanna without turning it into a desert. They rode horses in the scorching heat, cooked over a fire, and lived in minimal shelters shared with insects of every description. A few had some ranching experience, such as Ed Bass, who came from Texas and loved the new place. But most of the synergians had been living on the outskirts of town in New Mexico; now they found themselves out in the true bush, rounding up cattle from vast 20,000-acre paddocks. Stephen Storm recalled, "There were six or eight of us there, and we just swung into operating this station, which none of us really knew how to do. I mean some of us knew a little bit about horses, and some knew a little bit about cattle, and some knew a little bit about windmills, but none of us really had ever managed a quarter-of-a-million-acre station before, and so that was quite a dramatic first year. We experienced floods, and fires, and the whole range of things that can happen in the bush."[9]

As if that was not enough, the next year, in 1978, the Institute bought yet another five thousand acres of western Australian grassland—small in comparison, but still a huge chunk of property. They named it Savannah Systems and decided to focus there more intensively on managing the grassland. A first crew of arrivals quickly started setting up a minivillage of houses and gardens. Determined to restore the native grassland, they labored for hours that became days that became years, working by hand with adzes to hack out thickets of scrubby invading "wattle" (acacia) plants to make room for grasses to come back. They sowed drought-resistant plants, especially leguminous fodder crops, to help replenish the soil, and brought in cattle and horses to graze on unwanted weeds. And as at their other outposts, despite living on the fringes of civilization, the Ecotechnic group imagined their work to hold global symbolism. Here, in the remote Australian prairie, they would make "a model for the intelligent development of the savannahs worldwide," they wrote in a brochure about the project. Another site in the synergetic civilization was born.

The young ecotechnicians were having a great time. They were also reflecting their society's confused and passionate quest to work out the right relationship to the rest of nature, in a historical moment full of uncertainty over the subject. Multiple intellectual and emotional frameworks about nature coexisted and blended in the Ecotechnics members' minds. At times they pronounced themselves committed to resolving the perceived conflict between wilderness and technology; at other times they gave voice to old ideas about the possibility of mastering or wisely managing nature; and throughout, many remained captivated by romantic notions of the frontier, at a time when the planet seemed to be running out of frontiers left to explore.

The deep-rooted American fantasy of the frontier, the edge between civilization and wildness as a place to build one's character, was reborn in this new generation of rebels. Many who joined the Ecotechnic projects had traveled on personal quests all over South America, Africa, Asia, searching out their own frontiers. Ironically, through the Institute's projects, a group of Americans headed off to the remote Australian

Outback to become old-fashioned Wild West cowboys. The analogy was clear to participants like Stephen Storm, who came from Texas but never got to be a frontiersman until he helped found the Outback station of Quanbun Downs:

> I came along too late to participate in the American frontier. There were still people in my family alive who could tell me about what it was like on the frontier, and so I had all along kind of nursed this desire to experience what had happened in the building of America, and that was I think what Australia did for me, was it actually filled this gap, and so now I feel like I'm probably truly an American.[10]

The lure of the frontier mythos went beyond just the Americans in the group. Silke Schneider, an energetic young German woman, had worked an office job in the import-export business, hated it, and on an impulse left home to join a traveling circus with elephants and horses. On the road in the south of France, in 1980, she happened upon a performance by the Theater of All Possibilities. Meeting the actors afterward, she was entranced: "I was just really interested in where they all came from, because people would say, 'Oh, I come from Australia, I come from England, I come from Germany,' all these places." A year later, Silke joined their theater tour to America. "That's really what I wanted to do—to see the world." Upon arriving at Synergia Ranch, she declared in a Sunday night dinner table speech that she had always dreamed of riding the "authentic cowboy trails" through the American West. Everyone loved the idea, including John Allen. With their enthusiastic support, Silke—now "Safari" to her new companions—trained for two months riding horses at the ranch. Then, faithfully following the old cowboy routes, she set out on horseback from Santa Fe. She rode 1,200 miles, first east to Dodge City, Kansas, and then south to Texas. In classic theatrical form, she arrived in Fort Worth just in time for the opening of the theater group's new performing arts center, the Caravan of Dreams, at high noon. Soon after, she flew to Australia to the ranching station, still chasing dreams of the Wild West.

Why such fascination with the frontier? The Ecotechnics members were continuing to live out America's New World creation myth. In that myth, the American continent itself was imagined as a "new world" in which to live out ideals and dreams; within that continent, the frontier was imagined as the stage to act out that newness. However, the classic American frontier ideal contained at its root a great irony: the frontier mystique grew as the actual frontier itself disappeared. The more people chased the frontier, the less frontierlike it became. Historian Frederick Jackson Turner published his famous Frontier Thesis, arguing that the frontier had shaped the unique American character, in the 1890s—just as national censuses were demonstrating that the once vast wild frontier had mostly melted into defined tracts of land bounded by civilization. Thus the ecotechnicians, like many others, became post-frontier people obsessed with finding a frontier—as though they needed the illusion of a truly blank space on the map in order to do something new. "At last!" declared a character in a Theater of All Possibilities play, dreaming of heading to space with his friends: "*My* frontier! Where I'll find out if I'm a real man or not!"[11] Even as they tried to leave behind conventional society, the actors continued to play out the old creation story of North America.

As the network of projects thrived and grew, new members joined. Recruits showed up at the Ecotechnics outposts in the same haphazard ways that many had come to Synergia Ranch: through a friend or acquaintance, by watching a theater performance, by word of mouth. The decision-making structure spread out as the projects did; John Allen and other founding Ecotechnics members hopped around the world to the various outposts, while managers at each project made tactical decisions and presided at the dinner table in Allen's absence. But the old Synergia life structure remained the same. The project managers always formally presented themselves to the outside world as the Institute of Ecotechnics, but from within, the network of communes was known to members more simply as "the synergias." At each synergia, residents followed the lifestyle pioneered at Synergia Ranch: a steady schedule of nightly dinners, theater practices and performances, metaphysical readings and discussion, hard work on the land, and intense explorations of group dynamics. Their enormous agreed-upon inner aim was also the same as before: to grow as artists, scientists, and explorers; and by transforming themselves, to somehow transform the world.

Weekly theater and movement sessions remained part of the Institute, and each site's crew would generally stage at least one formal theater production per year. Productions ranged from comedies in which the actors made light of their own experiences, to other original plays by the perpetually touring Theater of All Possibilities company, to classic scripts from famous playwrights. The crew of the *Heraclitus*, as they sailed around the world, exchanged performances with local peoples wherever they docked. They wrote their own comedies and cabarets; they performed *Faust* in Singapore, *Hamlet* on the Red Sea. As they sailed up the Amazon, they stopped in indigenous communities, where they used gestures and dance to communicate with audiences. They also shot documentary footage about the native cultures they encountered along the way.

Of course, it wasn't all just a fantastic adventurous dream. Conflicts simmered in the various synergias, often linked to John Allen's quick temper. Said Ben Epperson, "I never got kicked out, I just quit four or five times and came back, but almost everybody was thrown out at one time or another . . . they would get pissed off, or John would get pissed off at them." Still, excitement pervaded, as young people, often with little life experience, suddenly found themselves working together to pioneer a new project in a foreign land. Recalled Phil Hawes, "One of the best things about it was John's knack for creating an interest on the part of the people in doing these various projects, and being able to manage the people flow. Say, if I was working on a project in Kathmandu and it was kind of winding down, and then John would say, 'Well, there's an opening in Australia or there's an opening in France. Which would you rather do?'"

Soon more than one hundred people were involved in Ecotechnics projects around the world, moving from one outpost to another in search of new adventures. By the mid-1980s, one hundred aspiring "sea people" had completed nine-month stints learning seamanship and diving aboard the *Heraclitus* as it sailed the world's oceans. Another hundred had apprenticed at the various land and theater projects, by John Allen's estimate.

The Institute of Ecotechnics' plans did not stop at wild nature. After all, much of the world was no longer wild; many of the world's people lived in cities. Therefore, John Allen and his cocreators reasoned, to become global citizens they also had to learn to work with the human habitat: the "urban megalopolis." To that end, using the same organizational structure (Bass as investor and a handful of managers on a corporate board), the Institute bought a three-story brick schoolhouse in the heart of London for their next big venture. They reworked the building to house an art space that they named the October Gallery, as well as a café, library, theater, and rooms for visiting artists and friends. Here their invented buzzword was "transvangardia"—an international mix of artists and thinkers, each one from "the living creative tip of whatever the art form is in that culture," said Kathelin Gray, the Theater of All Possibilities director who moved to London to help run the project. Though a small synergetic group lived together there, it was hardly a closed commune. The goal was "to interact with as many—not as many people as possible, but as many interesting people or creative people as possible," said Ben Epperson, who moved to the London house for a time. Dance, theater, and poetry performances drew new people in. The gallery walls displayed artists from Africa, Asia, the Middle East, and South America, as well as avant-garde European artists.

Finally, in 1982, the actors and builders began work on their ultimate temple to artistic creativity. This performing arts center they named the Caravan of Dreams. Its location was the unlikely cultural capital of Fort Worth, Texas—not quite the "transvangardia" scene of London, but it was Ed Bass's hometown, he had the money, and he was keen to help revitalize his city's deserted downtown, to fit in with urban renewal projects that his brothers were already bankrolling. The Caravan became the group's grandest creative center: a three-hundred-seat jazz nightclub on the ground floor; a theater, bar, and dance studio upstairs; a kitchen and living quarters to house the resident theater troupe and visiting artists. Crowning it all, on the rooftop shone the Desert Dome, a glass geodesic dome enclosing a botanical garden of dryland plants.

At last, in 1983, to complete the circle of projects, the Institute bought nine hundred acres of abandoned second-growth rainforest land, part of it formerly a coffee plantation, on steep hillsides in Puerto Rico. There they began another experiment in ecological restoration, planting rows of hardwood trees in the forest and committing to develop their own ecologically sound way to harvest timber. And so, by the early 1980s, the

Institute of Ecotechnics founders were reaching their dream: they had an outpost in nearly every major planetary ecosystem.

Once the Puerto Rico rainforest project was under way, a miniature world of sorts was complete. Ecotechnicians were at work in some of the most important and challenging biomes on the planet. Their effort to represent every piece of nature in their work was conscious. Synergia Ranch represented the world's deserts; the Australian ranches covered the grasslands. Hotel Vajra was perched in the mountains, the French farm estate in the temperate Mediterranean climate, the Puerto Rico project in the tropical rainforest, the London house in the urban heart, and, of course, the RV *Heraclitus* on the oceans. By working with all of these systems at once, they were trying to take on the whole world, piece by piece. As some of the founders wrote in their characteristically esoteric way, their goal was nothing less than "interrelating ecosystems, including man, his cultures, and his technosphere with the evolving biospheric totality on the planet earth."[12]

Was the Institute a network of innovative ecological and cultural projects, or a series of random international adventures bankrolled by an eccentric young billionaire—or some combination of the two? How could an eclectic collection of ecological restoration and theater projects, dotted around the world, actually have global significance in the way its members imagined? This paradox lay at the heart of the ecotechnicians' shared lives. As from the beginning, the group's aspirations and internal contradictions illuminated the spirit of their times.

As spaceships sent back the first images of the Earth from space, the environmental movement was taking on a planetary consciousness, portraying the round blue Earth in space as a single home. The *Whole Earth Catalog*, a collection of counterculture resources, was first published in 1968, and activists organized the first Earth Day in 1970. In the rising 1970s environmental movement, "Think Globally, Act Locally" became activists' mantra. Yet that slogan was incongruous to the core. "Acting locally" might lead to visible, tangible results in local ecosystems and communities. But as the new environmental movement embraced the image of the whole Earth, and problems of worldwide pollution and overpopulation dominated the discourse, the nagging question was obvious:

Would scattered groups of people "acting locally" ever be enough to soothe a troubled planet? By racing around the world from project to project, it was as if the Institute of Ecotechnics members were trying to solve that paradox for themselves, struggling to be local and global all at once. As though striving to convince themselves of their own global significance, the Institute's leaders made a point of representing the whole world in everything they did. The crew of the *Heraclitus* even aimed to log a theater performance on every continent, finally reaching that goal in 1989 with a production in Antarctica.

In its planet-embracing spirit, the Institute began hosting annual conferences at its outposts around the globe. Lubricated with food, drink, theater, and stellar lists of expert attendees, these were lively intellectual weekends; guest thinkers and synergia members came together to exchange lectures and conversation. The conference topics mirrored the group's aims to master the planet bit by bit. They began by addressing one biome at a time, then expanded astronomically. They started at home in New Mexico, holding a local Upper Rio Grande Ecology Conference in 1973. Next, year by year, came an Oceans Conference at their farmhouse in France, Deserts Conference in New Mexico, Mountain Conference in Nepal, Jungle Conference in Malaysia, and a conference in Australia on "eco-transition" zones between different ecosystems. Then, in a series of annual events at the farmhouse conference center in France, the themes and vision got bigger and bigger: a Planet Earth Conference in 1980, Solar System Conference in 1981, Galactic Conference in 1982, Cosmos Conference in 1983. It was as though the Institute organizers believed that after a decade of work and study, they had covered the earthly biomes and were ready to consider the stars.

Honored conference attendees jetted in from all over the world. The Institute of Ecotechnics directors invited their intellectual heroes from every field that might be necessary to the "synergetic civilization" of the future. Big-name guests over the years included biologists Richard Dawkins and Lynn Margulis, Buckminster Fuller, astronaut Rusty Schweickart, jazz musician Ornette Coleman. Guests gave talks during the day and at night enjoyed performances from the Theater of All Possibilities or other invitees. "IE conferences were explicitly designed to include scientists, artists, managers, and explorers and to include within science a range of disciplines," recalled chairman Mark Nelson. "One of the draws for the speakers/participants was the chance to get outside their own narrow field of study or action. None of the speakers ever received an honorarium, just economy air tickets."

Each symposium's balance of artists, scientists, and explorers reflected the hosts' belief that they must master all three realms—emotion, thought, and action. The Mountain Conference in Nepal exemplified this belief: American geologists lectured on mountain formation, plate tectonics, and soil; a Pueblo Indian spoke about sacred mountains; an anthropologist described the significance of mountains in world cultures; Himalayan mountain climbers recounted their adventures among the peaks. The combinations often led to amusing cross-disciplinary interchanges, as at the 1980 Planet Earth Conference. There anthropologists spoke about the history of humankind; various scientists lectured on the world's plants, oceans, geology, climate, and diseases; and Beat novelist William S. Burroughs delivered a rambling lecture on his personal vision of "The Four Horsemen of the Apocalypse," concluding that humans should transcend the beleaguered Earth and evolve to voyage space in an "astral dream body," sending some of the more orthodox scientists into fits while John Allen and Ed Bass respectfully asked questions of their counterculture hero.[13] Somehow all of these were parts of the knowledge the Ecotechnics members hoped to integrate and act on.

So the Ecotechnics members were playing with multiple models and archetypes of humans' role in nature—the mythic quest for the frontier, interbred with an increasingly global environmentalist consciousness. They personified their generation's symbolic, many-faceted, and sometimes contradiction-riddled quest to restore humans to their proper place in nature. In practice, Ecotechnic projects of clearing brush, herding cattle, and planting trees may have looked like simple restoration efforts. But meanwhile, multiple ways of thinking wrestled under the surface. The project members possessed an environmentalist attention to ecology, but they married that view to a strong faith in the power of human intelligence. The tension between those two frames of mind would one day animate Biosphere 2 as well.

This interplay of ideas reflected a deep uncertainty about the human place in nature, present throughout American ecological thinking at the time. The ecotechnicians' aims all boiled down to a single question, Mark Nelson told me years later: "Could we actually integrate this human system with the ecosystem so that both would benefit?" It was not a simple question. By the 1970s, the relationship between humans and nature had been thrown into confusion. The Institute of Ecotechnics members were dreaming of integrating ecology and human technology, in a time when the two seemed polar opposites. Average citizens lived more disconnected from the natural landscape than ever before, and it was not clear how to mend the rift. The first major U.S. environmental laws aimed mainly at stemming the damage, trying to keep humanity from mangling nature any further. The 1964 U.S. Wilderness Act, passed by Congress, protected nine million acres of federal land against human encroachment, defining wilderness as a place outside human influence, "where man himself is a visitor who does not remain."[14] Early environmental legislation, such as the Clean Air Act of 1963 and the Clean Water Act of 1977, focused on setting limits on industrial pollution. These key laws mostly aimed at constructing boundaries, making the line between humans and nature sharper, in order to keep nature safe. But where was a vision of how humans could actually live reintegrated into the rest of creation?

Could high-tech, postmodern humanity return to playing a more supportive, meaningful role in nature? For all the talk, this was still the question lurking within the Ecotechnics projects. The Theater of All Possibilities actors toyed with it in their own plays. In one original John Allen production, *Tin Can Man*, a surreal cast of characters plots to mastermind evolution on Earth. "Playing God!" exclaims one character. Another character corrects him: the goal is "acting as a cooperative agent of the gods."[15] Editor Stewart Brand had famously declared in the first *Whole Earth Catalog* in 1968, "We *are* as gods and might as well get good at it."[16] The synergians took that directive to heart.

The Ecotechnics members' more official writings and speeches reflected that aspiration to become "good" gods. In one such speech, Mark Nelson, the articulate young chairman, described the Institute's conceptual framework at the group's 1980 conference at their France farmhouse. An early arrival at Synergia Ranch, just out of college, Nelson had happily sweated in the daily labor of planting trees, building windbreaks, working with the soil, channeling storm runoff. At the conference ten years later, standing at a blackboard before scientists, anthropologists, and his friends, now he danced through abstract concepts to explain that work. He and John Allen had refined an intellectual system, he explained, to classify an "eco-region" according to "twelve categories, four inorganic, four organic and four superorganic": heat energy, orientation on the planet, atomic composition, and mineral structure; soil, species, groups, and eco-communities; individual human "culture-creators," schools, cultures, and "planetary decisions" affecting the area.[17] Nelson seemed to hope that if humans could organize information correctly, they could make intelligent choices about how to treat the land. The world did not need more quietly observed ecological data, but rather, a "science of management," he argued, describing the process of revegetating Synergia Ranch from desert to orchard: "At first you have to work and work and work if the system is headed downward, but when the system is reversed, starts to head the other way, nature begins to lend a hand, to organize to counter any downward trend to soil degradation . . . nature can act to further the evolution of the system."[18]

Such visions of humans aiding and guiding natural processes were unusual among 1970s environmentalists, but the Ecotechnic vision had deeply American cultural roots nonetheless. Many past American ecological thinkers and managers had believed that humans could intelligently manage nature. That thought pattern stretched from the work of landscape designer Frederick Law Olmsted, who replicated natural landscapes in Central Park and other urban spaces in the 1800s, to Aldo Leopold, the scientist-manager who pioneered wildlife management and worked on ecologically restoring his degraded farm in Wisconsin in the mid-twentieth century. These men of earlier eras loved the natural world, but had a strong faith that the human mind could help it develop.

Ecological theories are often about more than ecology; they can also be manifestations of deeply held beliefs about the purpose and potential of life. This, at least, was the case for the new ecotechnicians. For their work was about more than physical ecology—it was metaphysical. Indeed, the synergists seemed to believe (or hope) intensely that they could learn not just to manage nature, but to manage anything.

That faith flowed from John Allen. In his metaphysics lessons, he taught his companions to sketch diagrams of connected lines, called enneagrams, to order their thought processes for everything from writing a speech to managing an ecosystem to managing the ranch kitchen. Allen had built on this practice, which he had learned from John Bennett, the disciple of the mystic Gurdjieff. Going further still, John Allen wanted to distill logical systems of the world into a single body of numerical reasoning. *Everything* in the world could be managed, given the right mind and technique. In 1985, he would publish his complicated synthesis in a book called *Succeed: A Handbook on Structuring Managerial Thought*. In it, Allen argued that "this publication is the first time that these structures with their mighty power to organize the attention have ever been presented in a completely practical form":

> After Hegel and Kant, and Hegel's great students, Marx and Engels, the trail led to Pythagoras and his theory of the objective reality of numbers, Aristotle and his four causes, the Parmenidean monad, the Heraclitean dyad, the Buddhist five-term logic, Zen paradox (dyad), Buckminster Fuller with his triads, tetrads, and hexagons, Einstein's four dimensions, Beethoven's Quartets, the theological triad, projective geometry and equilibria at Mining Engineering School, Gurdjieff's Laws of three and seven, Ouspensky's six dimensions of recurrence, the Sufic Eight, Lawrence Durrell's Quartet and later his Quintet, and John Bennett's Systematics.[19]

Many who had been in the synergias insisted I read Allen's *Succeed* if I wanted to understand his thinking. I readily confess that I found the book's intellectual exercises incomprehensible. But most of the people recommending it had equal trouble explaining it to me in ways I could understand. Or maybe, I wondered, the point was that not everyone was supposed to be able to understand it.

Half a century before Synergia Ranch and the Institute of Ecotechnics, one of the synergians' heroes, the mystic leader Gurdjieff, had described the need for highly developed individuals to step forward and guide evolution in a world gone astray. In his immense trilogy *All and Everything*, Gurdjieff stated the goals of his philosophy and life's work: first, "to destroy in people everything that in their false representations appears to exist in reality"; second, "to prepare 'new constructional material'"; and finally, "to build a new world."[20] It was a poetic yet abstract directive, perhaps only a metaphoric instruction for seekers to alter their own minds and "build a new world" internally. But for the synergians, that process was happening in the material realm as well. With the synergia members' old lives left behind, and the Ecotechnic projects and way of life established in every planetary ecosystem, "new constructional material" had been readied. Now it was time to build the new world.

ACT II

The “space frame” rises into the air over the future biospherian farm area.

GENESIS

And the Lord God planted a garden eastward in Eden; and there he put the man whom he had formed.

And out of the ground made the Lord God to grow every tree that is pleasant to the sight, and good for food; the tree of life also in the midst of the garden, and the tree of knowledge of good and evil. . . .

And the Lord God took the man, and put him into the garden of Eden to dress it and keep it.

—Genesis 2:8–15

These are the generations of Noah: Noah was a just man and perfect in his generations, and Noah walked with God.

And God said unto Noah, the end of all flesh is come before me; for the earth is filled with violence through them; and behold, I will destroy them with the earth.

Make thee an ark of gopher wood. . . .

And of every living thing of all flesh, two of every sort shalt thou bring into the ark, to keep them alive with thee. . . .

—Genesis 6:9–19

EVERY WORLD HAS ITS CREATION STORY. IN 1982, AT THE INSTITUTE OF Ecotechnics Galactic Conference in southern France, Phil Hawes stood at the blackboard and held up a white sphere, not much larger than a basketball, for the guests to see. The gathering that year included assembled friends from synergias around the world, and the invited speaker-guests included a few world-famous icons of alternative thinking, among them Buckminster Fuller and microbiologist Lynn Margulis, champion of the Gaia Hypothesis. Now Hawes—who himself had bounced around the planet designing ecotechnic projects—stood before them all. From behind glasses and a long brown mustache, he explained his latest tiny creation, which he had made at John Allen's urging. It was a model of a spherical adobe enclosure to be built at a size of 110 feet in diameter. The group was by this time used to wild ideas, but the leap from the adobe houses that Hawes and his team had once built in New Mexico was ambitious even by their standards—for this sphere was designed for life in space. The spaceship would rotate in order to produce its own gravity, Hawes explained to the audience. Furthermore, it would be not just a vehicle, but a place to live, a self-sustaining world. Apartments, gardens, and wilderness areas would line the sphere's curving inner edges. Human inhabitants would cultivate plants and animals, including miniature cattle bred for space travel. Biodiversity would come from a store of cryogenically frozen genetic material for the propagation of twenty thousand different species. At the sphere's center would be a zero-gravity "Globe Theater" where actors and audience could float—in three dimensions, a true theater in the round. The plans were purposefully fanciful, a play of ideas, and no one, Hawes included, walked away with intentions of actually building such a structure. But it served another purpose: it got the conference delegates talking.

It was two years later, in the autumn of 1984, when Tony Burgess received a phone call at his home in Tucson that he would never forget. Tony was a red-bearded dryland botanist who had designed the desert garden on top of the Caravan of Dreams performing arts complex in Fort Worth. Now, two years after that project, Margret Augustine was on the telephone, and she wanted to know if she, John Allen, Ed Bass, and others could come meet at Tony's house at ten o'clock that Friday night. When they arrived and sat down, John Allen made an announcement: "We're going into the space race."

Two days later, the small group drove out of town along the base of the mountains, into the cactus-dotted hills near Oracle, to inspect the land they were already considering for their new project. John Allen had picked southern Arizona as the prime site for a world under glass, as year-round sunlight would keep plants growing in every season. Now

Allen wanted Tony and his friend and fellow ecologist Peter Warshall to make sure the proposed site had no major ecological problems—but, as Tony recalled, the deal was sealed when their party came over the crest of the hill and looked out at the breathtaking desert view. The setting was spectacular, under the foot of the sharp Santa Catalina Mountains. Here they could buy 1,800 acres of high-elevation desert. It was an open space of dried-up washes and low ridges, peppered with woody mesquite undergrowth, dry grasses, and cacti, and roamed by lizards and herds of wild peccaries. Only an hour's drive north of Tucson, the site sat within reach of two major airports, making it accessible to guests, yet far enough out in the desert to keep some privacy. The nearest neighbors lived several miles up the highway in Oracle, an isolated little town of working-class families and artists.

The site for the new world possessed only a little cluster of living spaces and meeting rooms that the Motorola Corporation, and later the University of Arizona, had been using as a conference center. The founders named their new land SunSpace Ranch. And only a few months later, in December 1984, the ranch opened for its first gathering: the Biospheres Conference.

Institute of Ecotechnics members again jetted in from their outposts around the world, with representatives present from the projects in Australia, Puerto Rico, Kathmandu, Santa Fe, London, Fort Worth, and at sea. The conference program listed the Ecotechnics members as "Research Associates" and John Allen as their "Total Systems Consultant"—their new professional roles in a theatrical science production. As always, the other invited guests were an eclectic and fascinating collection: space engineers, NASA astronauts, and visiting scientists in fields ranging from human medicine to chaos mathematics to microbiology. But this time they were not simply gathered for collegial intellectual exchange. They had been summoned to offer advice on a project that would need the expertise of everyone in the room. After dinner John Allen rose, and the lively chatter of the guests quickly fell silent. He greeted the gathering: "Friends and fellow students of the universe!" And he confidently told them their mission: to create "the first mitosis of Biosphere Planet Ocean"—the first offspring of their home planet, a new world. If this first experiment of Biosphere 2 worked, Allen proclaimed, new biospheres could one day support permanent settlements on other planets. With their help, Earthly life and intelligence could spread throughout the universe. They would start on this site in the Arizona desert, but their target would be, he concluded in patently grand language, "the objective history of man on this planet, which is the struggle to realize all possibilities."

The visiting scientists sat gaping as Allen announced that he intended to go to space for artistic reasons, not military reasons. "We were all stunned at that point," Tony Burgess recalled. "There was a lot of giggling. We couldn't believe this was happening." But, he said, "It was neat to see that they were going to use some interesting science to do something way out of the ordinary." After only a few days of talking, many of the scientists were hooked. After all, they were being offered the chance to play with millions of dollars, in order to recreate their favorite ecosystems of expertise under glass.

As they built excitement around their new project, the Institute of Ecotechnics crew already talked as if they were embarking on an expedition to a foreign land. When Mark Nelson, the chairman of the Institute and Allen's longtime collaborator, stepped to the microphone at the Biospheres Conference, his rhetoric rang less of biology than of manifest destiny, as he declared in a measured, slow, but strong voice,

> Perhaps childhood's end is at last at hand. Alexander the Great reportedly wept that there weren't other worlds to conquer. What would he say to the prospect of creating new worlds, and sending them into Cosmos? We stand as travelers at the crossroads, armed with technics so powerful as to both threaten the continuation of our species, and perhaps our very biosphere—but technics which also make possible the accomplishment of man's oldest dream . . . As children of Gaia, the Earth's present biosphere, we have the opportunity, unique in our history, and perhaps a necessity, to pay back the debt for our upraising, to initiate a new science and practice of biospherics.

His words were dramatic but fit the occasion. Later Nelson reflected, "Many of the people who were involved at NASA during the Apollo project of the sixties compared the spirit they found at Biosphere 2 with the excitement at NASA during those years—when everyone knew what was being attempted was damn near impossible but a hell of a challenge."

Some in the conference room already believed they could be heading for outer space within a few decades. (As a reporter later observed, "Here people talk very casually about retiring on Mars. Here people talk about Mars as if it were France."[1]) The last section of Mark Nelson and John Allen's little black book, *Space Biospheres*, first published in 1986 by the project's own Synergetic Press, would include sketches for a "Mars Science Station and Hospitality Base" made of four interlocking biospheres. The authors were already considering the mountains and canyons of the red planet for the most scenic place to build their next home. Their portrait photographs in the back of the book showed them stony-faced, wearing coats and ties, gazing intently toward the horizon as though toward the stars.

From the beginning, to its creators Biosphere 2 was much more than an ecological project, more than a glass greenhouse, and even more than a model space station. The structure's name evoked cosmic-size purposes; it was to be, quite literally, a new world. In the early 1980s, the term *biosphere* was scarcely in use, even among scientists. "Initially the name of the Biosphere company was Space Biospheres Ventures," recalled Kathy Dyhr, the project's publicity director, but "we just put it on our checks as 'SBV' because if you said 'biosphere' to anybody, even scientists would say, 'What the hell is that?'" The obscure term dated back to 1875, when an Austrian geologist named Edward Suess coined the word *biosphere* to refer to the envelope of life forms covering the Earth's surface. Then, in the early twentieth century, a Russian geochemist, Vladimir Vernadsky, had more fully elaborated the concept. Vernadsky explained in his 1926 monograph *The Biosphere* how all living organisms, functioning as a whole, together shaped conditions on Planet Earth. To Vernadsky, the term *biosphere* signified the totality of living beings, together with the air, minerals, and water that they controlled through biogeochemical cycles. Still, the Russian thinker's work remained little known outside the Iron Curtain. It would not even be published in English until Biosphere 2's builders put out a book of excerpts in 1986.

What would it mean, then, to create a Biosphere *Two*? The cocreators' logic was grand but poetically simple from the beginning. Every day, the

Earth's countless life forms were busy maintaining the chemical and ecological balance of their world, promoting conditions hospitable to life. It was only a matter of scientific tinkering, then, the Biosphere-builders reasoned, to make a second biosphere—to put together a combination of plants, animals, air, soil, water, and energy so that the constituent parts could evolve into their own self-perpetuating system. The Earth itself would no longer be the only known biosphere in the universe; it was, in the project's lingo, merely "Biosphere 1." The assembled scientists and builders of Biosphere 2 would be midwives to the birth of a new world.

Biospheres 1 and 2 would differ in obvious ways: Biosphere 2 would take up a giant glass greenhouse, not a whole planet; and Biosphere 2, unlike the Earth, would need a massive power plant to keep its air blowing, water flowing, and temperature stable. Yet Biosphere 2 would become *like* a whole world to its participants. In their manifesto *Space Biospheres*, John Allen and Mark Nelson set out their own working definition of a biosphere. They argued that a biosphere was a "stable, complex, adaptive, evolving life system," containing the five kingdoms of life and including multiple "biomes," such as the rainforest, ocean, and desert.[2] According to Allen and Nelson's logic, just as Biosphere 1 was open to inputs of energy from the sun, likewise Biosphere 2 should also be allowed to receive inputs of outside energy—even if those inputs came not only from the sun but also from a man-made power plant. Thus, with a few slippages and convenient analogies, the idea of Biosphere 2 came into being.

Though Biosphere 2 was barely a sketch on paper, in its creators' minds it was already the subject of many different dramas. The project leaders spoke of space, cosmic destiny. Meanwhile, the invited scientific consultants were excitedly talking among themselves about a different question: Did science have the ecological knowledge to actually build a new world? Many of the designers had experience in constructing ecosystems, but on smaller scales. Tony Burgess, a U.S. Geological Survey biologist, had previously put together the Desert Dome rooftop garden for the Caravan of Dreams arts center. Ghillean Prance, the Amazon expert who would plan out Biosphere 2's rainforest, had worked for years at the New York Botanical Garden, in its vast greenhouses enclosing exotic species. Walter Adey, the marine systems designer, was in the midst of building a miniature version of the Chesapeake Bay ecosystem in tanks in the basement of the Smithsonian in Washington, DC. But none of them had ever had to design a whole world that could survive on its own.

Carl Hodges of the University of Arizona's Environmental Research Laboratory would, along with his colleagues from U of A, oversee the development of Biosphere 2's agriculture systems and ecotechnologies. Hodges was an unconventional thinker and a doer, a well-known specialist in coaxing agricultural production out of harsh environments and in designing ecotechnological systems. And so, at that first Biospheres Conference in 1984, when he got up to make his keynote speech on the "The Do-ability of a New Biosphere," everyone was listening. Hodges proudly described his own lab's long track record of successful experiments: they had figured out how to grow crops with desalted seawater in the Middle East, how to farm shrimp in controlled enclosures in Mexico, and even how to use untreated seawater for irrigation of salt-loving plants. The key to "making the desert bloom," Hodges argued, was to learn from nature and base engineering designs on natural systems. If people designed a system smartly, they could survive on only about half an acre of land apiece, he estimated. Finally, with firm conviction, he made his pronouncement on the scientific prospect of creating an entire enclosed world: "It's totally doable."

The scientists eagerly discussed ideas for the huge new experiment. "We were going to use this as a focal point for all these disciplines," Tony Burgess told me years later, sitting back at his kitchen table in Tucson, his face lighting up as he recalled the initial excitement among his colleagues. "What we thought we were going to do . . . was that once we fused those disciplines, emerging from that project would be a whole new perspective on looking at and managing the ecosystems of the Earth." Tony himself felt that by setting up Biosphere 2's desert, he could repay his debt to scientists who had set up ecological observation plots in the Arizona desert eighty years earlier. Those plots had yielded research results that helped generations of scientists, including himself. "With this much focus on a small system, as [Biosphere 2 savanna designer] Peter Warshall pointed out, we were going to invent, and really refine, what he called 'invisible ecology,'" Tony recalled. "The ecology we naturalists had been trained in was the ecology of fuzzy things and trees and tangibles that you could see; this invisible ecology was going to be the flux of molecules, atoms, compounds, energy." The chance to just play with plants also appealed to the scientists. As a scientist working on his own PhD in the Arizona desert, Tony said, "I missed the creativity that one could do with more artistic things, and this project allowed me to see myself doing creative designs and construction . . . It was a very liberating experience." The grand experiment quickly became not just a consulting job but a passion for many of the hired scientists, he said. "It had to be a passion to get it done."

While the scientists dreamt of ecological possibilities, the Theater of All Possibilities members were imagining different plot lines. In their minds some were already headed for Mars. The script was already written in their plays. In 1983, the year before Biosphere 2 became a plan, the troupe had performed an original drama called *Kabuki Blues.* The play told the story of a supposedly fictitious guru figure named "Mr. Kabuki"—though his personality was coincidentally similar to the play's author Johnny Dolphin (John Allen), and the play's plot bore some striking parallels to the shared life story of its actors. It was a play about "the challenges of creating a viable yet noncommercial ensemble," recalled Kathelin, the theater director. "I love its wild and comic book plot that gets more and more surrealistic as the play progresses. It begins in a subway in New York and finishes running off with androids into outer space."

The actors published their play as a comic book, starring Mr. Kabuki's mischievous grinning face on nearly every page. The play begins in New York City, where the characters rehearse theater under Mr. Kabuki's demanding direction; when one cast member protests that the pace is too fast, whining to the guru, "This is an anarchist collective democracy! I want a vote!," Mr. Kabuki sternly casts him aside, asserting his dominance: "Art has to do with who can see." Suddenly, in act 2, the cast is transported to a ranch in "Catastrophe Flats, Australia," in the middle of the Outback, singing,

> Here we can say anything
> Here we can do what we need
> Here we can dance and sing
> Here we can plant our seed!

But the troupe does not remain content to live in isolated freedom. Mr. Kabuki shows them a telescope into the future, revealing the planet exploding in sex and violence, millions of babies being born and dying as bombs explode and pollution envelops the globe. In the last scene, the cast somehow ends up on Mars, traipsing toward a spaceship. Proclaims one, "Off we go to the galaxy to find ever new spirals of drama!" The final plate in the comic book depicts only a white castle in space, floating among the stars.

An advertisement circulated at the 1984 conference already sought out candidates for the privileged first eight expedition members, the crew of human biospherians. The sheet of paper read,

> Men and women between 18 and 30, who are scientifically disciplined and interested in both ecology and technics, in top health, emotionally open and willing to learn intensive group dynamics, able to work hard with zest for a common purpose, creative without having to be always the center of attention, and who wish to contribute to a truly historic project may apply.

All candidates were instructed, "To apply write a thousand word essay on your personal history; another on what you wish to accomplish with your life; and a third on human history, its potentialities for the next move."

At the same time that the core synergetic team were picturing themselves in an historic epic about space and evolution, they were enacting a very different scene for the rest of the world. On paper, the Biosphere 2 team was a network of highly professional corporations. John Allen set up the managerial structure of Biosphere 2 accordingly. The project would be a joint venture between two private companies, the first press releases announced: Decisions Investment Corporation (chaired by Ed Bass), which would provide the capital, and Decisions Team Limited (board members John Allen, Ed Bass, Marie Harding, Margret Augustine, Mark Nelson, and Kathelin Gray), which would provide the management. It was the same corporate structure that the core team had used before in their Puerto Rico rainforest project and in the Caravan of Dreams art complex in Texas. This time the synthesis would be a company called Space Biospheres Ventures. A glossy publicity brochure entitled "A Project to Create a Biosphere" explained in dry corporate prose, "The Institute of Ecotechnics (IE), an international ecological development firm chaired by Mark Nelson, presented the original concept of Biosphere 2 to DTL (Decision Team Limited) and has been engaged as project consultant for *total systems ecological management*." The brochure explained how the publishing company Synergetic Press would contract to publish books for the project; the architectural firm SARBID (Synergetic Architecture and Biotechnic Design) would draw the blueprints. Somehow the press releases neglected to mention that the intertwined boards and officers of these companies and organizations all included roughly the same people, and that the board members of Space Biospheres Ventures had also been living and working together in communal ecological projects and traveling theater troupes for two decades. The press releases also failed to note that Decisions Investment Corporation's capital all came from the personal accounts of young billionaire Ed Bass, whose fortunes were in the process of transforming from hundreds of millions to billions while his capitalist older brother invested the family's money in Disney.

The key players all took on new roles in the corporate production. Margret Augustine, the sharp people-manager, would be the CEO of Space Biospheres Ventures and coarchitect with Phil Hawes; as many observers later noted, Allen tended to be surrounded by powerful women. John Allen himself seemed unable to stick to just one title; starting out as "total systems consultant," he would later become director of research and development, then executive chairman. Alongside these titles, the list of institutions hired on to help design the Biosphere—New York Botanical Garden, Smithsonian Institution, University of Arizona Environmental Research Laboratory—made an impressive roster as well. And so while one play went on within the group of Biosphere-builders, another elaborate production was put on for the outside world, which gradually began to take interest in what the *London Times* identified as "an inspired group of scientists, architects, and other savants" working on a groundbreaking ecological project. "You read the accounts of the Biosphere, what you see in the press . . . they had no idea, it was just the surface," said Phil Hawes, who was busy sketching blueprints for Biosphere 2. "They had no idea what was going on really and the depth of the strangeness of it all." *Discover* magazine called the project the work of "one wealthy investor and a group of mainstream researchers" in an article entitled "The New World."[3] The magazine made a glowing pronouncement in 1987: Biosphere 2 was "the most exciting venture to be undertaken in the U.S. since President Kennedy launched us toward the moon."[4]

Such media decrees would take on an exaggerated appearance years down the road. But again, one of the most bizarre things about the Theater of All Possibilities' cosmic drama was how unbizarre it was, and how much, in its own fantastical way, it reflected other American dramas of its time. Popular culture seemed to have a taste for apocalypse and salvation. Science fiction movies and TV shows had already popularized the idea of eco-colonies in space as humanity's last refuge from an ailing Earth. The 1972 movie *Silent Running* told the story of a noble Forest Service agent in charge of saving the solar system's last forests on huge "forest ships" floating in outer space, since all forests had already died back on Earth. The 1982 television serial *The Starlost* chronicled life in the future on a huge "Earthship Ark" spaceship made of interlocking "biospheres." The eight-thousand-mile-long floating space colony carried representative remnants of all of Earth's cultures, after some unspecified "catastrophe of galactic proportions" that had destroyed the home planet in the year 2285. Thus Biosphere 2 was literally (though unintentionally) science fiction coming to life.

Such science fiction stories were remarkably imaginative—but more remarkable was how down-to-earth they actually were. In fact, they were not far off from the U.S. government's own plans. The year 1984 was not just the birth date of Biosphere 2 but the height of popular enthusiasm for the space race, a time when science fiction stories suddenly seemed to border on real possibilities. NASA's successful first launch of the airplane-like space shuttle in 1981 had changed the public perception of outer space. As one observer wrote, suddenly "[i]t had become possible to envision space as the province of some sort of transport vehicle (even looking vaguely like a commercial transport) which could carry anyone instead of the exclusive domain of supermen in capsules."[5] In the early 1980s, polls found that for the first time more Americans thought the U.S. government should "do more" in space than the number wanting the government to "do less."[6] In 1984 Ronald Reagan became the first presidential candidate to make space a major part of his election campaign, and promised he would direct NASA to develop a permanently manned space station within a decade. Civilians would have their place in the space future too, the government proclaimed, and so ten thousand American schoolteachers competed for the prized chance to fly on the exciting, highly publicized 1986 mission of a space shuttle named *Challenger.*

John Allen and his colleagues were not the only ones discussing space in the language of new worlds and frontiers. The space race, by the 1980s, had become the new vehicle for America's old New World creation myth. Ronald Reagan was plotting to launch missiles into space through his Strategic Defense Initiative (popularly known as the Star Wars program); but the U.S. government mainly sold the space race to 1980s taxpayers in the language of classic New World utopianism. "We must always be the New World," Reagan announced in speeches, repeatedly speaking of space as "the new frontier."[7] The 1986 report of the National Commission on Space, entitled *Pioneering the Space Frontier,* rhapsodized about American conquest "from the highlands of the Moon to the plains of Mars," as though celestial bodies were just an extension of the North American frontier:

> Five centuries after Columbus opened access to "The New World" we can initiate the settlement of worlds beyond our planet of birth. The promise of virgin lands and the opportunity to live in freedom brought our ancestors to the shores of North America . . . The settlement of North America and other continents was a prelude to humanity's greater challenge: the space frontier.[8]

In an invited speech at Biosphere 2 in 1989, NASA administrator Thomas Paine showed slides of distant planets and galaxies to the Biosphere builders and told them, "If you guys are interested in exploring, there's a lot of country out there."

In the mid-1980s, then, Biosphere 2 came at the right time. As a lush green alternative to NASA's mixed plans of space colonization and possible space warfare, it presented a more beautiful version of the government's own popular plans. Journalists eagerly played up the drama of a paradise in space. "Local scientists developing new kind of world near Oracle," the *Arizona Daily Star* headlines sang, announcing that

Biosphere 2 would demonstrate how "people, plants and other life forms could live indefinitely without support from the outside world"—and that similar biospheres "eventually might be used to colonize other planets."[9] Many onlookers assumed that somehow Biosphere 2 could save the Earth from some dreaded apocalypse. As one magazine journalist wrote, "Although no one is prepared to guess the outcome of their two-year experiment in cocooning, what they learn inside may help save us all."[10] Even the *New York Times* explained matter-of-factly that Biosphere 2 would "prepare for a day when Earth might be no longer able to support life—because of the collapse of the sun, perhaps, or more immediately a nuclear war."[11] To America, Biosphere 2 seemed simultaneously a potential utopia in space and an escape vehicle from the distressed Planet Earth—both a remade Garden of Eden and a Noah's Ark fleeing toward the stars. In the popular mind, Biosphere 2 embodied two extreme views of nature: an idealistic hope that people could live in total ecological harmony (inside Biosphere 2), combined with a fear that it might already be too late to achieve such cooperation on Earth itself.

Life in space was only one part of the Biosphere 2 creation story. An unspoken collective ideal of an ecological paradise also surfaced early in the design process. The glass structure would contain an intricately engineered rainforest, desert, ocean, savannah, marsh, and farm, featuring thousands of carefully selected species of plants, insects, animals—altogether, a miniature world. The metaphor of a "new world" resonated powerfully, both for the creators and for the general public. If a functioning space colony had been the only goal, Biosphere 2 could have been a much simpler beast, along the lines of the little controlled ecosystem experiments that both Soviet and NASA scientists had been constructing for years. Beginning in the 1960s, Soviet scientists had sealed themselves in airtight metal rooms for months on end in a series of experiments aimed at modeling long-term life in space. The first man stayed locked up alone in a hermetically sealed, tiny 4.5-cubic-meter cabin in Moscow, with nothing but chlorella algae to supply his air and part of his diet for thirty days. In Krasnoyarsk, Siberia, in one of ten experiments in a facility called Bios, a crew of three men stayed sealed up for 180 days with hydroponically grown crops. However, though these experiments recycled all their own air and water, they were not completely self-sufficient, importing some food and exporting human wastes. Meanwhile, NASA was also experimenting with a program it called CELSS (Controlled Ecological Life Support Systems). But as the name implied, life support systems was all that these projects were—pared-down living spaces packed with only a few superproductive food crops, heavily reliant on technologies to recycle air and water. The plants were just an extension of the machine. Biosphere 2, with its lush diversity of plants packed together under domes and pyramids, bespoke a much grander vision of nature.

There would be some practical reasons, of course, for Biosphere 2's abundant biodiversity. First was one of the most generally accepted rules of ecology: diversity brings stability in ecosystems. If some plants did not survive in the new glassed-in environment, the designers reasoned, then a variety of others would grow in to fill the gap. Thus the creators agreed on a strategy of purposely overpacking their world with a high number of species, allowing for extinctions along the way as nature navigated its own balance. Including a variety of different ecosystems would also help balance the gases in Biosphere 2's atmosphere. In winter, when most plant growth would slow down, the biospherians could "rain" on the dormant savanna to wake up the plants to produce more oxygen. Strategic rain on the desert could also extend that area's growing season; it helped that desert plants were accustomed to making their growth spurts during winter rains. Still, these scientific justifications for Biosphere 2's extravagant diversity only went so far. Why would Biosphere 2 count among its residents desert tortoises, blue lizards, and playful little African primates called galagos, commonly known as "bush babies"? Why include a fragile coral reef in the ocean, a waterfall gushing into a hidden pool in the rainforest, the sacred Amazonian hallucinogenic vine *ayahuasca*, or perfectly timed waves lapping against a sandy beach in the ocean? Practical considerations only told part of the story. Though they never stated it explicitly, the builders seemed to be creating an alternate earthly paradise.

By the second Biospheres Conference, in September 1985, the goal of creating a whole new world under glass was explicit. To kick off the gathering, Mark Nelson announced, "To create Biosphere 2 will give man his first opportunity to step into a new living world." The Ecotechnics

Biosphere 2's designers envisioned a true wilderness world under glass, as shown in this early design sketch.

SYNERGETIC PRESS.

team and their hired science consultants had returned to the conference rooms at SunSpace Ranch—this time not just for talk and inspiration, but for planning. One by one, each lead biome designer took the floor and explained how he planned to capture the essence of an earthly wilderness in the small space allotted. Tropical botanist Ghillean Prance described how his Biosphere 2 rainforest would consist of eight microhabitats representing different kinds of Amazonian terrain. In a "lowland rainforest," tall canopy trees cloaked in vines would reach for the sky; a meandering stream would evoke the Amazon's seasonally flooded riverbanks; a waterfall would gush into a pond; and in the center would stand a concrete mountain topped by a cool and damp "cloud forest" carpeted in plants on the microclimates of its cliff faces and terraces. All this would be crammed within a towering four-sided pyramid with a floor area of just 1,900 square meters. Tony Burgess, the desert designer, took a similar strategy. He was determined to squeeze all kinds of desert habitats into one square pyramid. He described how his team would construct a gravelly *bajada* slope like the alluvial fans of the Southwest, rocky (concrete) caves and cliffs containing a community of succulents growing on the cliff faces, a flat salt playa for special salt-loving plants, and even a tiny lone sand dune dotted with grasses.

Biosphere 2's aquatic designers would take a slightly different approach, announced Walter Adey, the marine design captain from the Smithsonian. Instead of picking out plants one by one, they would just pull in huge chunks of nature, put them together, and see what happened. "Just dig in," Adey urged his colleagues. "You can't hold back because you don't know . . . it's the grand adventure of the twenty-first century." Adey was already hatching a grand plan to create mangrove swamps in Biosphere 2. Out in the Everglades, he explained, marshes gradually transitioned from inland freshwater to oceanside saltwater zones, and therefore the assemblage of plant species would gradually shift along a three-hundred-mile watercourse to the sea. In Biosphere 2, he announced, he would jam all of this variety into a mere one hundred feet: dig up chunks of different kinds of marsh in Florida—whole mangrove trees, roots, muddy soil, and all—then truck them to Arizona, plop them next to each other in Biosphere 2, and watch what happened. The ocean would grow by a similar strategy, with sections of an entire Caribbean coral reef put together in a huge tank. The builders would collect corals from the sea, Adey explained, and place them in Biosphere 2 along with truckloads of seawater (plus huge quantities of Instant Ocean aquarium powder mixed with local well water).

Following the same vision of bottling up the whole planet, even the Biosphere's "Human Habitat," where the biospherians would live, played on the notion of creating a microcosm of the outer world. Architect Margret Augustine, at the designers' conference, described the high-tech habitat as a "microcity" or "micropolis." It would feature "different districts like you would find in any major city," she explained. The biospherians' eight apartments would be the "homes," the kitchen and dining room the "restaurant district." Computer and videoconferencing rooms would represent the "high-tech communications" sector, the chemical laboratories would be a zone of "hospitals and medical research and analytics," and an "industrial zone" in the basement would include a state-of-the-art machine shop to repair any broken equipment. Along the edges of the microcity, Augustine explained, would lie an agricultural "transition zone": a barnyard of goats, pigs, and chickens, and intensively planted, purely organic fields of vegetables and grain.

Finally, even the outer shell of Biosphere 2 would aim to represent the whole world, by echoing the greatest achievements of architectural history. Architect Phil Hawes, who had worked with Synergia Ranch's adobe in New Mexico, had all along been dreaming of "organic architecture," sketching huge, elaborate eco-castles of his own imagination. Now, at last, he had the freedom to turn his dreams into blueprints. His first sketches of Biosphere 2 showed soaring interconnected round domes topped by turrets and flags, linked with elevated bridges and walkways. The final form of Biosphere 2 would lack the flags, but would still end up looking something like the glass lovechild of an ancient castle interbred with a spaceship. Two matching step-pyramids would hold the desert and the rainforest, a set of three high-arched ceilings would encompass the agricultural fields, and space-age curving white walls would enclose the Human Habitat, under a bulbous white observation tower flying above it all. Despite the jumble of shapes, each piece of the architecture had its own significance, all designed to evoke a "synergetic civilization." Barrel-shaped arched ceilings were meant to recall one of the oldest architectural forms on Earth, dating back five thousand years—a form that had first entranced the Theater of All Possibilities actors on their caravan through Iran in the 1970s. Flat-topped step-pyramids would symbolize the temples and tombs of Mesoamerican and Egyptian empires. Last, to represent the indigenous tradition of the Southwest and the builders' own history as a team, the architects hoped to build the Human Habitat out of adobe clay bricks; but when they realized that making adobe airtight would prove nearly impossible, they settled on a shiny white metal shell. All of these shapes would be knit together in a geometric white web of steel triangles and diamonds, a "space frame," designed by Peter Pearce, an engineer who had worked with Buckminster Fuller and specialized in such constructions. The synergians had long believed they could synthesize the best knowledge of the past into a new, greater whole; their madly complex but strangely beautiful architecture would strive to do the same.

With all of these grand visions in the air, again the task of closing the second major design conference fell to Carl Hodges, the maverick ecoengineer, to inspire the team to move forward. "We are going to develop the most valuable two acres on the entire planet," he announced with great certainty. (In fact it would be 3.15 acres in the end.) After a year of conceptual visioning, Hodges declared, it was now time to get to work on the nuts and bolts of design—engineering, technology development, instrumentation—and test it all. He proposed a snappy schedule: build a much smaller minibiosphere in the next year to test the technology, start construction on the real thing a year later, build Biosphere 2 itself in a year and a half, lock it up for a few test closures, and get the humans inside with the doors sealed as soon as possible. Already the pace was quick. "We'll probably be building at the same time we're designing it," Hodges acknowledged, but proclaimed, "As Sophocles said, though you think you know the thing, you have no idea till you try it." As for the biospherians, whom he predicted would be inside their bubble in a mere four years (in fact, it would be six), Hodges exclaimed, "The only reason I can see that they should want to come out is to go to Mars—they've already been in Biosphere 1, they'll know Biosphere 2, so the only reason that they'd want to advance would be to go to Biosphere 3."

As Hodges spoke those words, doubtless some of the young biospherian hopefuls in the audience shivered in their seats. Just as in the early days of Synergia Ranch, seekers were again joining, coming from all over the globe on their own quests for meaning. Their personal stories reflected the many purposes already attached to Biosphere 2. Some felt drawn to the frontier of space. "I would like to be one of the pioneers going into space," Mark Van Thillo told an interviewer. The tall, lanky Belgian-born engineer was already an explorer of sorts. Vowing to see the world, he had traveled by himself around India, Asia, and Central America, then sailed around the planet on the *Heraclitus* for two and a half years. Now "Laser," as he was known in the synergias, said earnestly, he aspired to live in Biosphere 2, then go much farther: "Not actually settling Mars, but going on to the stars."[12]

Another aspiring explorer, German Silke Schneider, had ridden horseback through the American Southwest, traveled the world with the Theater of All Possibilities, and lived at the Ecotechnics projects in France and Australia. Now she transposed her frontier fantasy onto outer space.

"When I was little in Germany and watched cowboy movies, I kind of regretted that I had not been able to be in that old frontier, because I always thought that it was really exciting," she told an interviewer. And now, "Looking at the stars, I thought, this is where the new frontier is."[13]

One of the least likely new recruits was in his sixties: UCLA medical researcher Roy Walford, MD. Grinning from a gleaming bald head adorned with a long white mustache, he was lean but strong. His good physical shape was no coincidence, as his own body was one of his experiments. Roy was a well-known researcher in the medical science of aging. While the other Biosphere-builders hoped to be pioneers of space, he hoped to be a pioneer of time. Specifically, Roy believed he could extend the human lifespan up to 120 years through a proper low-calorie, nutrient-dense diet. Ever since his youth, Roy had wanted to find a way to extend the human lifespan; life just seemed too fascinating to be so brief. For the same reason, he periodically dropped his medical career for global adventures. Life would seem longer if one occasionally disrupted it to do something totally different, he believed. "I find it useful to punctuate time with dangerous and eccentric activities" he told a reporter.[14] Or, as he told me, "I find that periodically in order to preserve my sanity, I have to get out of the academic life . . . I like adventure." Early in his career, Roy and a friend used careful observation to outsmart casino roulette wheels, made a pile of money gambling, bought a boat, and sailed around the Caribbean for a year. Another time he hitchhiked across Africa with a girlfriend. Once he spent a year in India "wandering around kind of in a loincloth," he said, taking the rectal temperatures of meditating yogis to prove that they could physically change the state of their bodies by controlling their minds—and thus could prolong their own lives. And so, nearing his supposed retirement age, Roy came to Biosphere 2 for adventure. He had first met the Theater of All Possibilities actors in the 1970s and '80s, not through science but through art, as he was himself involved in experimental theater. Now he hoped to develop his own wild art and writing inside the Biosphere. But his job as crew doctor and medical researcher inside Biosphere 2 would prove to be his most vital role, for reasons no one had yet imagined.

A great number of the proud Biosphere 2 crew came with a different set of passionate motivations. For them it wasn't about just adventure and exploration. They wanted to heal the Earth. In Biosphere 2, these participants imagined, they could realize an environmentalist dream, by creating a place where at last people could live in total cooperation with all of the life forms around them.

For these ecologically minded Biosphere-builders, passion for the Earth and its creatures went deep. Fifteen years later, Abigail Alling still remembered with precision the first time she met the man who would change her life. A young graduate student with long hair streaked with blond, Alling—known to friends as "Gaie"—was enrolled at Yale, researching the vocalizations of blue whales for her dissertation. Growing up, she had spent summers on the Maine coast where her grandparents lived. There she fell in love with the sea, and, she recalled, "It was there that at the age of seventeen I decided to sail the world's oceans." But as a graduate student in environmental science, Gaie felt overwhelmed by the evidence of ecological destruction everywhere she went. In 1985, coincidence struck; while she was flying to one of her whale research sites in Sri Lanka, Gaie and John Allen happened to be on the same plane. He was flying into town to meet up with the *Heraclitus,* which was in the middle of sailing around the world. Bright student and future mentor sat and talked over tea in the port of Colombo, and John Allen described the Biosphere project to her, scribbling tiny sketches on a napkin. Taking in his rapid-fire conversation and electric eyes, Gaie was hooked. Biosphere 2 "offered a brilliant solution of how to harmonize ecology and technology . . . This was new! This was exciting!" she thought. "Biosphere 2 just floored me. I didn't really know what he was talking about, except he was talking about putting all this miniature world under glass in the Arizona desert, and didn't I maybe want to help with the marine environments, and I remember thinking this was the most amazing, extraordinary—everything that came out of John's mouth was just fascinating. It was just like, who *is* this person, and where did he come from, and who are all these people doing all these incredible, amazing things?"

By the time John finished telling Gaie about Biosphere 2, the *Heraclitus* came into port and dropped anchor. She felt awed at the first sight of its unique appearance, its black hull and huge sails. Shortly after, she dropped out of graduate school and applied for a job at Biosphere 2.

Linda Leigh, who would become a biospherian living alongside Gaie, told a similar, intensely personal story of environmentalist despair and hope. She grew up in Wisconsin, a nature-lover. In the 1970s, at the same time Synergia Ranch had been founded, Linda "moved to the mountains to be a hippie and protest and so forth," she said. Later, making her way through the world as a working scientist, she studied wolves in the forests of Washington state and collected plants on the Alaskan Peninsula. By 1984, Linda was already living in Tucson, working as a consultant for The Nature Conservancy, evaluating lands to protect for endangered plant species. What oceans were to Gaie Alling, the land was to Linda Leigh; thoughtful and concerned, she felt intimately connected to the natural world. When she heard about the planned Biosphere 2, she too felt irresistibly drawn in. The enthusiasm of the synergetic group out in the desert energized her. At that point, working in environmental science, Linda felt jaded and frustrated with the slow pace of progress and the rapidity of destruction. She would always remember a conversation with Norberto Alvarez-Romo, a member of the Biosphere 2 management, as a pivotal moment: "Norberto Alvarez-Romo said to me, 'Linda, you can't do this yourself'—save the planet, or conservation of ecosystems . . . That had a big impact on me. So when I saw this group of people who were so dedicated, and had such an incredible goal—there was this motivation to go forward, no matter what, and accomplish this goal—I said, 'Yeah, count me in.'"

She quit her job in Tucson, moved out to the Biosphere 2 site, and jumped into the small crew living and working together. Linda had been earning $100 a day at her job in Tucson. Now she only made $500 a month (plus free food and cheap on-site lodging) at Biosphere 2. In reality, she acknowledged, at the start she "barely knew what the project was about, I have to admit; *barely* knew what it was about." But "the idea was we were all working towards this magnificent thing. And there was some kind of value in that," she said. "I agreed with the goal they were working on, which was to learn about how the Earth works; in my opinion that was what the goal was."

What I found most fascinating about these stories, however, was what they left out. No one told me exactly how Biosphere 2 would help the Earth, but that hope was in the air around the project. Just before entering the Biosphere, Linda would tell an interviewer, "I feel a very strong debt to the Earth, and to the world as we know it. It's a debt that I would like to pay back, and I think I can pay that back in the work that I do through Biosphere 2, and the information that we can get out from Biosphere 2 to Biosphere 1."[15] Journalists likewise perpetuated the belief that the biospherians would save the Earth: "If they get bored, they can console themselves with the thought that even though this grand voyage of discovery may not be leaving the Earth, they could bring back knowledge that will help mankind escape extinction," wrote the *London Times*.[16] Still, no one agreed on exactly how Biosphere 2 would rescue Biosphere 1 from environmental crises. Early news reports announced that Biosphere 2 could act as a model for future enclosed refuges to save endangered species. The project's own press releases asserted that Biosphere 2 would foster the development of new "biospheric science" and sophisticated "ecotechnologies," much as NASA had long argued that the space program would generate "spin-off technologies" for use on the home planet. Scientists involved in Biosphere 2 hoped that it could serve as a model for ecological restoration efforts, as rainforest designer Ghillean Prance wrote: "Can we use it as a demonstration of how the despoiled and abandoned areas of Amazonia and the other rainforests can be treated? Can we use it as a showcase to show the alternative to destruction?"[17]

But the exact environmental applications of Biosphere 2 were never easy to pin down. Perhaps it was because Biosphere 2 was not solely a physical experiment. The pristine miniature wilderness was also a highly symbolic construction. Many working on Biosphere 2 believed that their creation's symbolic value would wake up the public and "set in motion deep-seated cultural changes," recalled Tony Burgess. By seeing the enclosed, interdependent world of Biosphere 2, the creators hoped, Earth's people would realize that they too lived in a finite biosphere, and that they had better learn to take care of it. "I really got intrigued by the idea that making a biosphere could revolutionize the way people thought about the environment, because I was always an activist," said Kathy Dyhr, the Biosphere 2 public relations director. "Everything that's sustaining the lives of those creatures is within that Biosphere, which is the first thing that if someone really wants to really have an ecological perspective they have to get their arms around . . . It was the most revolutionary environmental teaching tool that I could think of. I thought John was brilliant to come up with the idea." Kathy tried to emphasize in press releases that Biosphere 2 would help rescue the Earth not only as a "laboratory to study ecological processes," but also to help humans "learn how to do a better job of stewardship of Biosphere 1."[18]

John Allen was a master of grand pronouncements. Biosphere 2 would "catch the imagination of humanity," he proclaimed at one conference on the site. In his own colorful book on the Biosphere project, he wrote, "Biosphere 2, scientific model, symbol, affirmation, helps us understand life, and therefore, ourselves."[19]

Why aspire so fervently to rebuild all of nature under glass? Ecological justifications for Biosphere 2's astounding biodiversity told part of the story. But the project's metaphysical resonance also drew on one of humanity's oldest and most powerful symbols of paradise: the garden.

The very word *paradise* comes from the Persian *pardes*, which meant "enclosure" and signified a garden in ancient Mesopotamia. The first paradise, the first symbol of utopia for Judeo-Christian and Islamic civilization, was the Garden of Eden. Over the course of centuries, various societies and empires designed their gardens to give life to their concepts of an ideal world. In the process—as would hold true at Biosphere 2 as well, in complex ways—the garden designers often embedded in their creations not just their perceptions of order in nature, but also their own ideas of power and social order. Islamic Moghul emperors built highly ordered gardens, divided into four quadrants to represent four rivers issuing forth from Eden as described in the Bible; each neatly symmetrical garden was a controlled, ordered miniature universe to be commanded and arranged. Medieval European gardens mimicked the shape of rectilinear, walled cloisters, embodying the belief that a simple, obedient, churchly life provided the route to heaven. During the Italian Renaissance, gardens became more expansive, often featuring a house on a high vista point, integrated into a garden fanning outward, which in turn blended into the surrounding landscape—emphasizing a belief in the goodness of creation, with humanity, its crowning jewel, at the center of a harmonious whole.

By the age of European exploration and conquest of the globe, gardens began to reflect a new consciousness of the vast world abroad. As before, official gardens became places where elites could act out their power, symbolically, in a silent world of plants. Now imperial gardeners developed the tradition of trying to represent the whole world in a single garden—a trend that Biosphere 2 would follow four hundred years later. In sixteenth-century Europe, royal gardeners created symmetrical, meticulously ordered landscapes intended to represent the entire globe. Official gardens in Padua, Italy; Oxford, England; and Paris, with its famous *Jardin du Roi* (Garden of the King), consisted of four symmetrical quadrants, each one planted with the carefully arranged and catalogued vegetation of one of the four known continents: Europe, Asia, Africa, and America. Through these gardens, European rulers presented themselves as masters of a huge but highly ordered world.

That impulse to represent the entire world lived on in modern gardens. By the twentieth century, as one historian of gardening observed, the one trait unifying all European and American gardens was that they were "based on designs borrowed from every period of European garden design," as well as other gardening traditions from around the world.[20] This eclectic tradition found expression in places like the Duke Gardens of Somerville, New Jersey, a lavish set of eleven greenhouses begun by the wealthy conservation philanthropist Doris Duke before World War II. Its eleven gardens included a lush Italian-style courtyard; a formal French garden ringed by ivy-draped lattices; an "American desert" featuring giant spiny cacti; gardens in the styles of China, Japan, and even the Moghul emperors; and a tropical rainforest—altogether advertised as a "mini tour of the planet earth," a chance to "walk around the world in about ¼ mile."[21]

By the late twentieth century, the impulse to represent the whole world in miniature extended beyond mere gardening projects. Such efforts spanned a diversity of settings and human groups. In an odd coincidence, the resulting megaprojects often involved some sort of geodesic dome. One such dome sat at the center of Auroville, a solar-powered international village, founded in South India in 1968 as an experiment in "human unity." There European expatriates and Indian villagers came together in an opening ceremony to put handfuls of soil from 121 countries into an urn at the center of their future town. They planned to devote an entire quadrant of the town to an "International Zone" with pavilions representing great cultures around the world. The symbolic representation of the entire world was central to the community's identity as a new world unto itself. A few decades later, in 1982, on the other side of the world, Disney opened its EPCOT center ("Experimental

Prototype Community of Tomorrow"), displaying a strikingly similar effort to represent the entire Earth in one place. Walt Disney had originally envisioned EPCOT as an actual utopian living community under a climate-controlled dome. The $1.4 billion theme park that resulted instead aimed to represent the entire planet, containing a "Future World" full of futuristic technologies and a "World Showcase" featuring eleven pavilions representing different countries—including a Mexican pyramid, Norwegian village, Italian plaza, and Japanese pagoda. Thus, by the late twentieth century, a range of visionaries, from spiritual commune-dwellers to mainstream entertainment magnates, were striving to create miniature representations of the whole world. Moreover, their grand creations bespoke a grand hope: to make their small worlds more harmonious than the world outside.

Biosphere 2 would soon be mirrored in two other enclosed ecological projects, each aiming to incorporate all of the world's ecosystems under a single roof. One was the Montreal Biodome, a zoolike structure that would open to the public in 1992 (less than a year after Biosphere 2's completion). The Biodome, like Biosphere 2, strove to represent the whole planet, featuring a tropical rainforest, a Canadian coastline landscape, a northern forest where biologists even simulated the seasons to make trees lose their leaves, and a "Polar World" containing one of the largest penguin populations outside of Antarctica. Across the ocean, scientists later designed the massive Eden Project in Cornwall, England, as an icon for public education about conservation. Featuring five thousand different plant species from tropical, subtropical, and temperate climates, arranged under round geodesic domes, the Eden Project would begin admitting paying tourists in 2001.

Such global models, like Biosphere 2, seemed to express the hope that at least on a small scale, everyone and everything in the world could come together and get along. But why has Western culture fixated for so long on creating these symbolic paradise gardens? Though the latest megagardens have become increasingly scientific and modern, they grow from mythic roots. For the name of one of the most recent scientific gardens-under-glass gives a clue to the deeper meaning of these landscaping efforts: Eden.

Indeed, the world-encompassing gardens of the past five hundred years might be considered as scenes in a single ongoing, evolving Eden project. As historian John Prest has argued, the creation of royal botanical gardens in Europe was largely a response to the discovery of the Americas and the consequent realization that the Garden of Eden was not going to be found on some undiscovered corner of Earth; humans would have to recreate Eden themselves. Analyzing the writings of European imperial gardeners, Prest found that they believed that by assembling all the world's plants in one place, they were fulfilling a divine imperative to bring together the pieces that had been lost when man fell from innocence and Eden was destroyed: "Now that God had revealed the hitherto withheld part of the creation [America], men could go a long way towards recreating the Garden of Eden by gathering the scattered pieces of the jigsaw together in one place into an epitome or encyclopaedia of creation, just like the first Garden of Eden had been."[22]

Separated by nearly five hundred years, these European gardeners and the modern-day Biosphere-builders shared striking parallels, down to the precise language they used. The Theater of All Possibilities reworked the same mythic imagery to tell their own stories. Even when the image of a lost Eden remained unnamed, it lingered below the surface, as in a passage from the actors' adaptation of Goethe's *Faust*—which they published the same year that they gathered for the first Biospheres Conference in 1984. The play proclaimed,

> Doom! Doom!
> You have destroyed
> A beautiful world . . .
> We carry its scattered fragments
> Into the void . . .
> Magician,
> Mightiest of men,
> Raise your world more splendid than before . . .
> Build it up again![23]

"The metaphor of the Garden of Eden was never explicit, but nonetheless inescapable," Tony Burgess, the magician in charge of the desert, would reflect later. "We would return to the Garden, return to grace with Gaia, through biospheric gates."[24]

Maybe this was why young environmental scientists, desperate to heal the destruction of the Earth, felt so drawn to Biosphere 2. The women who would take care of Biosphere 2's land and sea hoped passionately that by creating their own biosphere and living in cooperation with it, they would also be healing the outside world around them—correcting humanity's fall from ecological grace by building a new, harmonious garden. "We've done a lot to Biosphere 1 . . . we've changed it," Linda Leigh told one interviewer. "Biosphere 2 is to change it in better ways."[25] To Gaie Alling, manager of the Biosphere 2 ocean, Biosphere 2 would be "a re-creation, using technology, to bring back what we may lose, and that is life."[26] Even biospherians devoted to the more technological side of the project loved the magic of creation. Mark "Laser" Van Thillo, who would live inside Biosphere 2 as manager of technical systems, said, "Somehow in maintenance, or mechanical and electrical systems, I feel like a magician . . . Just like flipping on that switch, and bang, there is light."[27]

The American media played up Biosphere 2's resonance with Eden. *USA Today* called the project a "man-made Eden." The Biosphere's creators sometimes drew on that symbolism themselves; the back cover of a book by two biospherians, *Life Under Glass*, quoted journalists who called the project "Planet in a bottle. Eden revisited." The story of Noah's Ark cropped up with equal frequency in written materials surrounding the project. The Biosphere Press published a children's book entitled *The Glass Ark: The Story of Biosphere 2*. In 1984 the Theater of All Possibilities published John Allen's adaptation of the ancient Mesopotamian epic of Gilgamesh. In *Gilgamesh*, in a story similar to Noah's, a hero with a boat saves all of life during a world-destroying flood. Many journalists also sought to use the Noah's Ark myth in order to make sense of Biosphere 2. A video documentary about Biosphere 2 called it "Noah's Ark for Mars?" Various news articles referred to a "greenhouse ark" (*Seattle Post-Intelligencer*), "landlocked ark" (*Arizona Daily Star*), "Noah's Ark—The Sequel" (*Newsweek*), and "a band of latter-day Noahs" (*Washington Post*). Such mixed metaphors showed the continuing imaginative power of old mythic forms, even as uncertainty remained over whether Biosphere 2 was a return to paradise—an Eden—or simply an ark, an escape vehicle to preserve species from disaster. Noah's story tells of a world in which most humans have become evil, bringing destruction on themselves in the form of a flood sent by God—but a few humans remain good, rescue the world's creatures, and ride out the disaster until they can start life over. The plot strikingly matched what some of the aspiring biospherians hoped to do in an interplanetary context.

It might appear surprising, on first glance, that the stories of Genesis should resonate so strongly during the creation of Biosphere 2, among a group of people who were hardly Bible readers. But the decidedly secular builders were not the only ecologists of their time to find meaning in biblical myths. In the late twentieth century, many voices in the growing environmental movement drew strongly on the age-old drama of Genesis—even in totally nonreligious contexts. In a seminal 1967 article in *Science*, historian Lynn White argued that Western civilization's mistreatment of nature was rooted in the biblical creation story, in which God gave humanity dominion over all living things.[28] The article, "The Historical Roots of Our Ecologic Crisis," remained widely cited for decades afterward. White had struck a nerve with his suggestion that Westerners were somehow doomed forever to live out biblical creation myths, even if they no longer literally believed in them. Broader environmental campaigns also drew on biblical imagery. Early awareness-building campaigns to save the Amazon rainforest played on caricatures of the Amazon as a last pure Eden, as historian Candace Slater has demonstrated by examining media and publicity materials.[29] As Evan Eisenberg wrote in *The Ecology of Eden*, assessing modern attitudes toward nature, "If you insert a probe into any body of environmental thought, you will find, somewhere near its heart, a firm if amorphous idea about Eden."[30]

The Eden story held a message, in the eyes of both Judeo-Christians and environmentalists: originally entrusted with care of the Earth, humans had sinned and fallen away from the natural state of paradise. For centuries, various thinkers have interpreted the Eden story to imply that humans must use their power to recreate that lost paradise and regain their state of grace. Historian Carolyn Merchant unearths this trend in a variety of contexts in her book *Reinventing Eden: The Fate of Nature in Western Culture*.[31] Merchant traces echoes of the Eden myth throughout Western history: in the settlement of America as the New World, in the Romantic landscape paintings of the nineteenth century, and even in the beginnings of modern science and radical environmentalism. In all those endeavors, a few individuals believed that they could bring society back into its proper relationship to nature. Many early European scientists explicitly viewed their experiments as a way of reversing humanity's fall from the Garden. Francis Bacon, generally accepted as one of the fathers of modern experimental science, wrote in 1620, "Man by the Fall, fell at the same time from his state of innocency and from his dominion over Creation. Both of these losses can in this life be in some part repaired; the former by religion and faith, the latter by arts and science."[32]

Some of the Biosphere-builders—specifically John Allen and his covisionary Mark Nelson—have protested against my suggestion that they were trying to build a "perfect new world." Indeed, that was never the stated aim of Biosphere 2. Yet as an outsider, combing through fragments from the project's past, I had to wonder if on a deep, perhaps unconscious level, the team of scientists and synergians were working to rebuild a mythical lost world, to restore humanity's lost role as keepers of nature.

To build that world, recruits culled from the Institute of Ecotechnics projects began streaming in from all over Biosphere 1. They moved their possessions and lives to the little cluster of apartments on SunSpace Ranch. There the core team members again joined in the collective lifestyle of the synergias. Weeks followed the same packed schedule as at earlier Ecotechnics projects: weekend theater practices, Sunday night speeches, intensive group self-examination work using Bion's theories of group dynamics and John Allen's complicated thought-structuring exercises, and of course, nightly dinners together. In one new twist, Tuesday nights' old theatrical literary dinners took on a scientific theme, as the actors studied up for their new roles as biospheric ecologists. Tuesday evening's assigned readings now came from scientific texts like Eugene Odum's *Fundamentals of Ecology* and Lynn Margulis's work on microbiology and Gaia. Everyone still came to special dinners in costume as a character from the reading, but now they might arrive dressed as a bacterium or a ray of sunlight.

Some members of the building group had lived this patterned life for a decade and a half; others were encountering the lifestyle for the first time. "I had no idea what I was getting into by joining a group of people in a synergia," said Linda Leigh, the conservation biologist who originally joined the project for its environmental goals. "I found out once I was living there that I had to be in the theater performances, that I had to be in the meditation . . . that I had to participate in all of the dinners, and all of the after-dinner activities, giving speeches and so forth." But she found the theater and group dynamics work surprisingly fulfilling, she said. "We were working on ourselves to be able to think in a way that we would be able to use to build a Biosphere . . . We studied groups intensively, and how groups work, and we applied that to how we were doing the Biosphere." Work on Biosphere 2 still loosely followed a Gurdjieffian philosophical model, she said: "Your work on yourself is for yourself. Your work with the group is for the group. And then your work on the Biosphere is for the universe." Meanwhile, outside the inner group of around thirty people who lived, ate, and acted together at the project site, many of the other employees and scientists had little idea exactly what was going on after hours.

At the same time, the hired ecologists who commuted to Biosphere 2 were busy inventing their own science of creation. The "biome design captains" met repeatedly and drew up their wish lists of species. Even though they spoke in professional academic language, they were searching for a way to understand what "wild nature" is, in order to recreate it faithfully. As I leafed through records of the scientists' quandaries and decisions, I found that they seemed to be grappling with unsettled questions inherent in the contemporary concept of nature itself. The old myths of Noah's Ark and Eden lingered below the surface, but those biblical stories never provided fully adequate ways of conceptualizing nature for such a complex scientific undertaking as Biosphere 2. Yet science, it seemed, had not yet conceived a more suitable modern myth to explain and prescribe humanity's role in nature.

On the most basic level, the design team approached Biosphere 2 as an ark. Much as Noah had represented all created animals in his ship, the Biosphere scientists determined to pack in as many species as possible. The builders agreed to seal the doors of Biosphere 2 for one hundred years, as though to make a totally clean break with the old world of Biosphere 1. "We were all absolutely convinced we were closing for one hundred years. No question," said Linda Leigh, who coordinated species lists for the terrestrial "wilderness."

But integrating all organisms under one roof would be tough in the face of modern understandings of ecological interconnectedness. "Noah's Ark is a metaphor, not a schedule," complained savanna designer Peter Warshall, who wanted more time to build a world gradually.[33] Somehow Noah had magically squeezed all the world's animals onto a boat three hundred cubits long, fifty cubits wide, and thirty cubits high, according to the Bible—a volume roughly equivalent to 2.3 million cubic feet (a "cubit" being around 20.6 modern inches). Biosphere 2 would be nearly three times as big, with a volume of six million cubic feet, but the scientists knew that they would not have nearly enough space to include everything. Ghillean Prance, the British-born rainforest design captain, framed the challenge to his colleagues at the 1985 conference: There were at least thirty thousand different vascular plants in the Amazon rainforest where he had spent his life working, and one hundred thousand in rainforests all over the world, plus anywhere from three million to thirty million kinds of rainforest insects; how could he choose only a few hundred species? The plants that made it inside the ark would have to be the ones most useful to the biospherians or to the ecosystem as a whole, he reasoned. Bananas, guavas, passionfruit, breadfruit, ginger, starfruit, lemongrass, coffee, and other edibles would get to grow in the rainforest to supplement the biospherians' harvests from their agricultural fields. Other plants made the list because they would play important roles in creating a structurally authentic rainforest: dominant canopy trees (but species not too tall, as Prance worried they could shoot up to hit the glass ceiling ninety feet above); fast-growing pioneer trees to create shade while the slower trees were still coming up; epiphytes and vines hanging; understory ferns and shrubs; ground-cover plants; mosses. Prance even proposed including a seed bank, possibly cryogenically frozen, in case the rainforest trees did not naturally reproduce—the Noah's Ark all-or-nothing approach demanded an analogy to Noah's pairs of animals for breeding.

Still, Prance told the conference gathering of his colleagues, he worried about whether all these plants would be enough to create an authentic rainforest. His version of ecology was more complex than Noah's, and he was trying to think up every possible interaction between his species and their environment. What if too much sun came in through the glass walls and made it too bright for some rainforest plants, which liked to grow in the shade underneath a dense canopy? (To deal with this possibility, he proposed a dense sheltering belt of ginger and banana trees to surround the rainforest's edges.) Furthermore, what if plants accustomed to growing near the equator, where days were almost the same length all year long, refused to flower at the higher latitude of Arizona as the seasons changed?

Choosing Biosphere 2's animal residents presented similar dilemmas. One by one, favorite animals were selected, their human advocates duking it out over the final list as they constructed a "master food web" connected with imaginary lines. The agricultural consultants started hunting for the ideal miniature chicken, pig, and goat breeds that would give the human biospherians some food but not eat too much themselves. In the "wilderness" areas, all the ecological considerations attached to each animal made the selection process increasingly complicated. Tony Burgess wanted desert tortoises—and ordered the soil scientists to include only soft sand so the tortoises would not cut their feet. Peter Warshall, the savanna designer, ordered a low wall to be built along the cliffs on the savanna's edge because he feared that the tortoises, which had never seen cliffs before, might walk off the edge and fall into the ocean. Warshall wanted termites in the savanna, but the engineers were afraid they might chew through the caulking sealing the glass windowpanes together, leading to a series of "termite taste tests" to make sure that would not be a problem. Warshall was also in charge of selecting one species of hummingbird out of more than 390 found in nature, ruling out those with overly specialized curved bills and those with complex mating dances that might send them crashing into the glass.

Using early versions of email, the designers created an ecosystem by electronically shooting ideas back and forth across the country from their home institutions. Walter Adey, masterminding the marine systems, had rights to design the stream running through the savanna. Walter belonged to the extreme shovel-everything-in school of thought, and late in the game he suddenly slipped a weasel into his official wish list for the stream. Peter Warshall, who was trying to carefully coordinate the food web, protested that a weasel would screw everything up. (No weasels were included in the end.) "It was like people had personalized their biomes," recalled Linda Leigh, who had the job of helping all the different wilderness specialists communicate with each other.

John Allen insisted on galagos, commonly known as "bush babies," to provide company for their fellow primates, the humans. The reason: one of his idols and acquaintances, the eccentric Beat writer William S. Burroughs, had become enamored of tree-dwelling lemurs during his travels in South America and suggested that Biosphere 2 include something similar (lemurs proved a little too large for the miniature rainforest). Finally, an entomologist from the Bishop Museum in Hawai'i, Scott Miller, was flown in to pick out the insects, a crucial job because insects would be necessary to pollinate rainforest plants. The tricky part was that many rainforest plants had evolved with specific insect pollinators in the wild. One tree might require a species of bee that had the right shape and strength to push into its flowers and become coated with pollen, but that bee in turn might require a specific set of flowers from other plants to sustain the bee's life cycle through the rest of the year. Some of the wilderness designers wanted bats to pollinate their plants, and they hired a bat specialist to pick the best species, but then the numbers came in: each bat would need to eat twenty moths per night, and for that to happen, statistically each bat would need to encounter at least two hundred moths per night—so bats were out.

The designing biologists' minds led them in circles; the more information they uncovered about each species, the more complications arose to present new headaches. "I became amazed at how little was known," recalled Silke Schneider, who worked on preparing the domestic animals and insects for the Biosphere. "How little was known about which lizard, what time of the day and at what height of the tree does he feed, and how big can the bug be that he eats, and the bug people saying 'we don't know.'" The biologists showed an infinite capacity to get consumed by the details. They were supposed to be designing an entire world—yet, said Linda Leigh, "we had people who were thinking, 'Oh God, what if this lizard dies?'" Coordinating between the ecologists and the Biosphere's engineers was also a challenge, she said. "How big do you have to have the screening over the air intake valve so that you won't kill all of the moths that are two centimeters in length that the lizards are going to depend on?" It was hard just to get engineers and ecologists to learn to speak each others' languages. Following the species-by-species ark approach, Biosphere 2 already seemed too big for any one person's brain.

But even as they dove into precise biological details, the scientists still seemed attached to a more mystical concept of nature: nature as *wild*, and in that wildness, somehow sacred. They were constructing a totally artificial world, yet they excitedly debated how to make Biosphere 2 take on the imagined qualities of wilderness. The designers consistently referred to their rainforest, desert, savanna, marsh, and ocean as the "wilderness" areas. Their obsession with creating a true wild feeling in Biosphere 2 suggested that on some level Biosphere 2 would be a temple. That sense of its sacredness would be deeply connected to the idea of wilderness. The Biosphere 2 rainforest could not be just any rainforest; it must represent *the* rainforest, the wildest place on Earth. Linda Leigh described the rainforest design process as a deeply aesthetic experience: "When collecting the plant species for the Biosphere over the past few years, I found that the sensations became important," she wrote. "Not just to get the right plants in, but to get the right feeling, the right sensation. Is there a slight breeze on your cheek in the place where there should be a breeze on your cheek? Are there a lot of little microhabitats where you can walk from one place to another and feel a five to ten degree change in temperature, which gives more habitat for organisms?"[34]

Linda's aesthetic sense, in this case, was connected to physical needs; the right temperature variations would mean the right habitat for different creatures. But such practical concerns could not explain all of the attention to aesthetics, as in the Biosphere 2 desert, where Tony Burgess also pursued a vision of an idealized wild landscape. "The essence of a desert is vast expanses of nothingness, and the real challenge is to catch that feeling of spaciousness, and abstract it, and encapsulate it in a square 120 feet on a side," he told an interviewer. He wanted to create "a harmonious setting that only a very trained eye would be able to tell is a product of artifice rather than nature."[35] Along the desert's edge, separating it from the savanna grasses, would grow a tangled, weird, and wonderful belt of thorny plants from around the world, designed to overwhelm the senses—for the desert was also meant, with its strong scents, intense heat, and spiky forms, to be "a place for extremes," Burgess said. He tried to imagine how a biospherian would feel walking out from this dense thornscrub into the open desert pyramid, hoping the contrast would make his baby desert feel wide and vast as it would in nature—even though it would cover less than fifteen thousand square feet.

By the time that Biosphere 2 came into being, pure wilderness had long since become modern environmentalists' version of Eden. That view of wilderness-as-paradise was itself an unnatural concept, however—a product of modern environmentalists' love of nature, but ironically, equally a product of their society's disconnection from wild nature. In the Bible, "wilderness" had originally possessed negative connotations as a desert, a barren and fearful place, the very opposite of the tame, fertile paradise of Eden. That revulsion at wilderness carried over into the settlement of North America, where Pilgrims and others tried to carve a society out of a "hideous and desolate wilderness," as one arriving on the *Mayflower* put it.[36] But in the following centuries, the wilderness concept turned on its head. The more wilderness itself began to disappear, the more it became seen as sacred. In the early twentieth century, conservation crusader John Muir described nature as "terrestrial manifestations of God" and declared wild forests to be "God's First Temples."[37] The transition to wilderness-as-holy was the product of a time when humans lived farther from wilderness than ever before, as described by historian Roderick Nash:

> [W]ilderness has evolved from an earthly hell to a peaceful sanctuary where happy visitors can join John Muir and John Denver in drawing near to divinity. Such a perspective would have been absolutely incomprehensible to, for example, a Puritan in New England in the 1650s.[38]

By the late twentieth century, the wild *was* its own sort of God, holy in the minds of its defenders. As long-time Sierra Club president and ecological activist David Brower wrote, "To me, God and Nature are synonymous."[39]

Of course, the Biosphere-builders' quest to create their own "wilderness" was riddled with paradoxes. Repeatedly the designers ran up against a core dilemma: real "wilderness," in their minds, thrived on chaos, unpredictability, and disturbance; Biosphere 2 would be too planned to be truly wild. The typical savanna "is so into being disturbed that you cannot even think of a savanna except as a disturbance community,"

Peter Warshall stated in a speech at the 1985 Biospheres Conference. Savannas in nature are highly resilient ecosystems; unlike rainforests, they are accustomed to climatic extremes, Warshall explained to his fellow scientists. African savannas actually depend on the nibbling of grazing animals to stimulate new growth, and fires are essential there to clear out woody vegetation and maintain the balance between woody plants and grasses. The most important skill for a savanna manager to have, Warshall argued, would be knowing just when to burn the grassland to stimulate new growth. Since Biosphere 2 would contain neither browsing animals nor fires, the biospherians would have to take on the role of master disturbers, aggressively cutting back thick grasses and woody stems, which would prove a Herculean task.

The Biosphere 2 desert too would depend on stress, Tony Burgess told his colleagues at the same conference. Scarcity of water defines a desert, he pointed out. Knowing just how far to stress the desert and then when to relieve it with "rain" from sprinklers would prove the key managerial tactic; the biospherians would have to watch each plant's life cycle to know when watering would stimulate maximum growth.

Burgess's and Warshall's idealized, highly disturbed little desert and savanna did not superficially resemble the grand ordered gardens of European history. But for them and their generation of ecologists, chaos and disturbance were the new version of holy order. Changes in ecological thinking paralleled broader changes in contemporary philosophy, as the concept of complexity uprooted any lingering faith that the universe possessed some neat natural order. Early in the twentieth century, ecologists like Fredric Clements, who pioneered the concept of succession, had seen nature as orderly and progress-oriented. Clements and generations of his followers believed that ecosystems evolved toward a steady, stable "climax" state. But by the late twentieth century, that sense of progress was gone. "Nature, many have come to believe, is *fundamentally* erratic, discontinuous, and unpredictable," wrote environmental historian Donald Worster.[40]

Biosphere 2's too-calm air would be another lack-of-disturbance problem. In the wild, normally winds would pollinate savanna grasses and spread their seeds; but Biosphere 2 would lack even a breeze. Artificial fans would prove a poor replacement, as they could only blow seeds a few feet. The designers toyed with the idea of putting an air blower running along a track up and down the length of the savanna, but this solution proved too expensive (even for the extravagant construction budget) and still would not have distributed seeds in a truly random way. The scientists also feared that in Biosphere 2's still air, big plants would grow weak from the lack of resistance. Indeed, many trees would later sag and bend over in the windless rainforest, eventually requiring ropes and pulleys to hold them up. The toppling trees became a visual metaphor for many life forms in and out of Biosphere 2: many organisms seem to crave the disturbance that nature provides, and even need disturbance to prompt them to grow strong. The complex role the biospherians would have to play to provide missing ecosystem functions inside Biosphere 2 soon became clear, in the words of one Biosphere design conference participant: "They have to be pollinators, they have to be herbivores, they have to be fire, they have to be wind and rain."

Adding to the contradictions, though Biosphere 2 was supposed to be "wild," it was also supposed to have a rainforest, coral reef, and desert magically get along under one roof—something that would never happen in such a small space outside in nature. The designers did their best to guard their own little wildernesses from each other. The rainforest would require humidity, but desert plants liked dry air; giving in to this reality, Tony Burgess finally decided to model his desert on coastal deserts found in places like Baja California, where cacti grew adapted to foggy weather. Likewise, Ghillean Prance wanted to hold humidity in his rainforest and protect it from the salt spray of the seawater below; he planned a thick belt of bamboo along the rainforest's southern edge to shield it from the rest of Biosphere 2. Walter Adey, the marine man, worried about his coral reef, one of the most sensitive systems in the Biosphere. In Biosphere 1, where oceans covered 70 percent of the planet, the seas' immense volume helped dilute pollution; runoff from streams would barely affect water chemistry far out in the ocean. But in Biosphere 2, the incoming stream water would contain so many nutrients that it could promote too much algae growth in the ocean, suffocating the corals. For this reason, Adey planned a living "scrubber" system to pump the ocean and marsh water through a series of trays and screens where growing algae would suck nutrients out of the water—a complicated setup, which, like so many of the other wilderness-protection tasks, would cause the biospherians hours of worry and tough physical labor down the line.

Biosphere 2's planned wilderness was thus supremely un-wild. Indeed, it manifested the contradictions inherent in the very concept of wilderness at the end of the twentieth century. In 1964, the Wilderness Act, the single greatest nature preservation law in U.S. history, designated vast tracts of federal land as "wilderness," never to be developed or used for human purposes other than recreation. The act defined wilderness, "in contrast with those areas where man and his own works dominate the landscape . . . as an area where the earth and its community of life are untrammeled by man."[41] However, there was something paradoxical about land becoming "wild" through an act of law. As one observer put it, in answer to the question "What is wilderness?," "wilderness is managed land, protected by three-hundred-page manuals specifying what can and cannot be done on it."[42] As wild lands were sealed off under human law, they became somehow more tame; modern "wilderness" was, at bottom, a human creation.

Biosphere 2 was born of a bipolar culture, obsessed with both the wildest nature and the most sophisticated technology. And so it was fitting, adding the final paradox to the creation of Biosphere 2's wilderness, that the entire ecosystem would be unable to survive without a basement full of machines. Biospherian Roy Walford would later call it "the Garden of Eden on top of an aircraft carrier." While the ecologists were busy designing Eden, a team of engineers was working on its life support system. Without the geological and climatic processes of the Earth, many "natural" cycles would have to be replaced by technology in Biosphere 2. The water cycle was a prime example. Originally the designers had hoped water would condense on the glass walls and magically rain down on the plants inside Biosphere 2, but eventually they settled on placing controlled sprinklers throughout the wilderness. Large fans would keep air circulating within and among the biomes. A wave-making machine, driven by a vacuum pump, would create turbulence in the ocean, necessary to keep the corals happy. Water cooling towers next to Biosphere 2 would provide the air-conditioning; chilled water would flow through sealed pipes underneath the wilderness, absorbing heat, then flowing back out. This system would crucially guard against the scorching temperatures possible in a desert greenhouse. If the cooling system failed, head engineer William Dempster calculated, the Biosphere 2 wilderness would cook at 150 degrees Fahrenheit on a sunny Arizona day. Finally, to keep all of these cycles running, Biosphere 2 would require huge amounts of energy: up to three megawatts at any time, though two megawatts would be the normal load. Originally the designers had hoped to use solar power, but backed off when they realized they would need acres of solar panels on the surrounding hillsides, at a cost of tens of millions of dollars. They settled on building their own power plant out in back of the Biosphere, which would run on a combination of natural gas and diesel.

Biosphere 2, even in the design phase, was demonstrating just how much support humans unthinkingly receive from the rest of nature on a daily basis. Ecological economists have tried to illustrate the incredible value of the Earth's environment by trying to calculate the dollar value of the "ecosystem services" that nature provides for free, but Biosphere 2 made the point more vividly, as William Mitsch, editor of the journal *Ecological Engineering*, later observed:

> Biosphere 2 operational costs and annualized construction costs, on the order of $10 million per year, allow us to estimate the real cost of recreating planet Earth ecosystems on the order of ~$\$10^9$ km^{-2} $year^{-1}$ if we had to do it from scratch. The ecological message of Biosphere 2 is clear—we should appreciate and try to understand the ecosystems in the Biosphere that we have.[43]

However, the Biosphere 2 designers saw no contradiction in creating a wilderness that was also a high-tech machine. In fact, they reveled in it. Biospherian Mark Nelson jokingly called it "Frankensystem or Alice in Ecoland" in a poem he wrote while inside.[44] John Allen liked to call his brainchild "the cyclotron of the life sciences." As project engineer William Dempster put it, "The cyclotron was a transformational apparatus in the science of physics that enabled entirely new kinds of investigation of atomic particles. Biosphere 2 is a transformational apparatus that enables entirely new kinds of investigation of ecosystems."[45] The fluid boundary between life and technology was reflected in the names of various parts of Biosphere 2. The elaborate network of computerized sensors, which would monitor temperatures and gases throughout the wilderness, became known as the Biosphere's "nerve system." Two white geodesic domes, to sit like flying saucers alongside Biosphere 2, became known as its "lungs." The two lung chambers, designed by William

Dempster, would inflate and deflate as air pressure changed inside the Biosphere, to keep the wilderness's glass walls from shattering as air expanded and contracted with changes in temperature. The mix of bodily metaphors revealed Biosphere 2's hybrid nature—a true cyborg.

Thus Biosphere 2 was born with multiple identities. "If you ask twenty people who were part of the project what the aim of it was, you would receive close to twenty different responses," biospherian Jane Poynter later observed.[46] If Biosphere 2 seemed to harbor contradictory identities, perhaps the schizophrenia was only a reflection of attitudes toward nature back in Biosphere 1. The glassed-in world embodied the thought patterns of the society that built it. From the Egyptian-style pyramids recast in steel bars, to the use of metaphors like "Glass Ark," to the mechanical "lungs," images of past and future constantly collided. The project captivated participants and spectators with the powerful image of a "new world"—but even this aspiration to create a "new world" was centuries old. But Biosphere 2's reliance on old myths was not the product of some hidden nostalgia or backwardness; it was evidence of a search for a meaningful role for humanity in nature. Myths have always played a valuable role in human societies. As Karen Armstrong wrote in her *Short History of Myth*, "We are meaning-seeking creatures . . . human beings fall easily into despair, and from the beginning we invented stories that enabled us to place our lives in a larger setting, that revealed an underlying pattern, and gave us a sense that, against all the depressing and chaotic evidence to the contrary, life had meaning and value."[47] Biosphere 2 arose in a postmodern time short on meaningful myths. Humanity had accumulated heaps of scientific information about the natural world, yet at the same time this highly informed society was causing more and more ecological crises. Americans enjoyed all sorts of technology but lived in less contact with the wild than ever before. Rational-minded, postreligious citizens possessed no single agreed-upon belief system to guide their relationships to the rest of nature. Biosphere 2, for its builders, countered that emptiness; they were giving birth to a new creation mythology for confusing times. And so this "new world," like the old world, was a grandly symbolic, highly contested terrain, even before it was born.

THE POWER OF LIFE

You have given me all that I prayed for,
Sublime spirit.
You revealed Nature's splendid secrets,
And gave me power to hold and enjoy her . . .
All living creatures have you linked to me,
From sky and water and forest . . .
But now I see, man's given nothing perfect.

—*Faust,* Theater of All Possibilities adaptation, 1984

What we call Man's power over Nature turns out to be a power exercised by some men over others with Nature as its instrument.

—C. S. Lewis, *The Abolition of Man,* 1943

THE THEATER PRODUCTION BEGAN IN EARNEST WITH THE CONSTRUCTION of the set. Out of the ground sprang the buildings that would be necessary just to prepare for Biosphere 2. First came a series of research greenhouses where candidate biospherians began testing out candidate plants for inclusion in their new home. The scientific design captains and teams of helpers from the Institute of Ecotechnics crew began roaming the country and the world—Guyana, Australia, Mexico, the Caribbean—on collecting expeditions. The new greenhouses in Arizona would hold the plants they brought back until Biosphere 2 was ready.

A cluster of small windowless quarantine sheds also quickly went up in order to fumigate the arriving plants against pests and diseases. Next came a hive-like cluster of low metal rooms known as the "insectary," to house collected insect colonies; an analytical chemistry laboratory;

and office buildings to hold the creative team and their increasing numbers of subcontractors. The centerpiece of these offices, perched on the hillside, was a shapely, two-story white building with high arched ceilings and windows looking out onto the Biosphere 2 construction site, a stone's throw away. This was Mission Control.

On a clear desert winter day in January 1987, work on the new world began at last. Standing in a ring on a white line in the bumpy, cactus-covered ground, marking out the future outline of Biosphere 2, a team of Space Biospheres Ventures employees gathered to ceremonially break ground. Margret Augustine stood at the center, holding a gold-painted trowel in one hand and a champagne glass in the other. Consultant Carl Hodges, standing next to her, gave the call over the megaphone, and everyone bent over with their own little painted shovels and started to dig. The *Tucson Citizen* proclaimed,

> While most of us went about our usual tasks yesterday, an event took place 35 miles north of Tucson that could be more important than the discovery of America.
>
> Ground was broken for Biosphere II.
>
> It could be nothing short of the first step to creating permanent life on another planet—if not the discovery of a New World, at least the making of one.[1]

Bulldozers dug the rest of the three-acre hole in the ground. Teams of workers lined Biosphere 2's future basement with a bottom layer of stainless steel, like a ship's hull sealing it off from the earth of Arizona, and poured cement on top of the metal.

But as in past Ecotechnics projects, the key raw material would be the intense energy of commitment from passionate people. Scientists came and went from the site daily, but the core team of Biosphere-builders, assembled from Ecotechnics projects around the world, lived on campus around the clock. Many of the key participants traveled on missions all over the world to collect species or consult with experts. But when they were there in Arizona, Biosphere 2 became their life. They ate dinner together daily and spent long hours into the night sitting around tables hashing out ideas about their miniworld. John Allen compared the structure to "the army, in mining camps, in lumber camps, in agricultural camps, in expeditions, on adventures, in construction camps in remote areas, on ships, in all of which I've spent a lot of time . . . to me, Biosphere 2 was a construction plus expedition camp." For the core team living the synergetic lifestyle at the site, work and life became inseparable.

There was so much to do that each member of the team would get to take charge of his or her own area of Biosphere 2. Sally Silverstone, known to her friends as "Sierra," flew in from the Ecotechnics rainforest project in Puerto Rico to join the Biosphere 2 team. When she found that she was one of the few group members with any bookkeeping experience, she began looking after the project's finances. Soon she found herself financial controller of the company, in charge of signing off on dozens, then hundreds of contracts, as millions of dollars began passing from Ed Bass's Decisions Investment Corporation into the project site. Sally also began apprenticing to Phil Hawes in the design studio, where he showed her for the first time how to do architectural drafting. Soon she was coordinating between engineers and a studio of twenty-two hired draftsmen as they furiously sketched the head visionaries' dreams into design plans for the Biosphere and its numerous support buildings.

The learning on the job went right to the top of the organization. Margret Augustine, who had started out spearheading the small but dedicated *Heraclitus* construction crew in the early 1970s, became coarchitect (along with Phil Hawes) and CEO of the entire project. Her highest education consisted of a BA in architecture awarded by her own companions at the Institute of Ecotechnics (one of the few degrees the Institute granted) and three years of "self-financed world travel studying cultures and cities," according to her résumé.[2]

And so began the builders' biggest theater production of all: learning to play the roles they were already living. Kathy Dyhr, who had first moved to Synergia Ranch in the 1970s, had no formal background in media work; soon, as one of the most charismatic and outspoken team members, she was the Biosphere 2 director of public affairs. At first she put out a few press releases detailing the project. Within a few years, as word of Biosphere 2 spread, inquiries started pouring into her office from curious reporters—first from Arizona papers, then from TV stations and magazines all over the country and even the international press. She had never done anything like this before—but, she reasoned,

"Of course, no one had ever designed a Biosphere before. It was kind of like right then and there, everybody knew the job description was going to have to be made up as you went." And, she said, "that was the kind of people it was—the kind of people who would be challenged by that, and intrigued by that, and interested by that, and not totally cowed by the idea." The builders had adapted before to new adventures, whether at sea, on horseback, or on stages from London to Texas to Nigeria; this was just the newest role. Everyday life became a sort of theater performance, recalled Linda Leigh. "The idea was that . . . everything is an illusion, everything is theater; the Biosphere is some of the most amazing theater that there is, so we'd better be good at being good actors."

Even the Biosphere's construction managers learned as they went. Kathy Dyhr's husband, Bernd Zabel, had come to Synergia Ranch in the 1970s as a young idealistic German in search of the American utopian communal lifestyle. Now he found himself cosupervisor of construction on a $150 million project. His cosupervisor, Laser, had gone to technical school in his native Belgium, but neither one of them had much background in building. "I had no experience in construction, and I was made the general manager of construction," Bernd recalled matter-of-factly. "In order to be general manager of construction, I had to play theater. I had to play as if I knew what a construction manager were doing." But, he insisted, it was not daunting; it was exhilarating:

> The beauty of this whole Institute was you were often thrown into situations where you had to swim; you had to do it . . . Some people they put in the wrong position, they failed miserably, washed off, got kicked off. Other people did OK. It was an experiment. But for me, for example, for me it was great; I never would have had the experience to be a general manager of construction for a hundred and fifty million dollar project . . . Johnny was really good at raising the energy level of this whole thing . . . There was a saying we had, which is, 'Take your heart, throw it out in front of you, and run to catch it.' It was like you really throw yourself far out, far beyond what you were thinking you were capable to do. It was exciting.

The Biosphere continued to rise from its foundations. Dozens of workers and contractors were hired from the local area—most of whom had more construction experience than their bosses. "Those guys didn't know anything about construction or standards or code. It was such a joke," said Rod Carender, a local construction worker who started out maintaining the greenhouses and then helped build Biosphere 2. "If it was your day to be the construction manager, you were," quipped Jeff Boggie, a local construction worker who joined the project in 1985 to put in the Biosphere's electrical systems and came to know every wire and switch inside Biosphere 2. But still, their bosses' group dynamic was impressive, Boggie said. "What first intrigued me was how well they seemed to work together . . . I'd never seen people really work together that well." The construction workers were curious about their quirky supervisors, the likes of whom a small town like Oracle, Arizona, had never seen. "I come from mining, working on mines all my life; I've worked underground seven mines in five states, I've been eight thousand feet below the ground, working construction as an iron worker," Rod Carender said. "But I never did anything like this—research, science, all this stuff was great, I loved it. I loved coming here and doing this, it was totally out of the ordinary."

More than a decade later, Rod, one of the last old-timers working at Biosphere 2 for Columbia University, would look around to make sure no one was watching, then drive me in a golf cart out into a little dry creek bed, tucked behind one of the old labs, to talk about the past. Parking hidden in the bushes, he excitedly told me stories of the people he had worked with: how no one told the construction workers anything about the cosmic aims of their bosses; how at first they were locked out of the construction site on weekends so the future biospherians could rehearse their plays in private. Years later, Rod still hung onto a bunch of photos of Theater of All Possibilities actors in wild costumes, which he'd found in a dumpster.

For those a little closer to the inner circle, the atmosphere was more exciting. "Everybody was really open and you could go up to people and talk to them, and if you showed an interest in working stuff, you were immediately drawn in. It was really dynamic," recalled Gary Hudman, a computer techie who was hired in 1988 by Hewlett Packard to work on the Biosphere's massive computer control network. Later he became so engaged in the project that he moved to an apartment on site and began working directly for Space Biospheres Ventures. However, Hudman said, "They were trying to do a lot of things really fast . . . They had a lot of young people in place, they didn't have a lot of experience, hardly any education. They were really enthusiastic, but they were making a lot of big mistakes." John Allen would title his glossy coffee-table book about Biosphere 2 *The Human Experiment*, ostensibly a reference to the two years the biospherians would spend inside. But outside Biosphere 2, a massive human experiment was already under way.

John Allen did not plan to enter the enclosure himself, but he did become the first official human guinea pig. While Biosphere 2's construction was just beginning, a much smaller baby biosphere was nearly finished just a few hundred yards away: the Test Module. The module would be a cube-shaped, one-room glass house packed with plants, just big enough for a single person to eat, sleep, and tend a few crops. The Test Module would aim to test out every aspect of the Biosphere project: the caulking used to seal leaks between the glass panels; the use of an expandable metal "lung" chamber to accommodate changes in air pressure; the data collection systems; the human team of ecologists and engineers working together; and the very proposition that a closed system of soil, plants, and water would survive and support human life. In September 1988, John Allen entered the Test Module for the first brief enclosure. He tended the plants and wrote in his journal for three days while a team outside carefully monitored the air chemistry and his vital signs. He emerged, radiant, and proclaimed, "I knew with my body as well as my intellect and emotions that Darwin and Vernadsky were right about the power of the force of life."[3]

That faith in "the power of the force of life" was driving the birth of Biosphere 2. Here Allen sat in a luxuriantly green, self-contained tiny world of his and his friends' making, with their mammoth dreamed-of "new world" rising out of the ground not far away. Everything had come true so far; the dream of a self-sustaining world seemed realizable. If this test module full of plants could sustain one man, surely a three-acre biosphere could provide a comfortable home for eight.

Still, with Biosphere 2 under construction, the question remained: would it all work? No one could say for sure. The faith that "the power of the force of life" would sustain Biosphere 2 did have a solid scientific basis. The evidence came from successful closed ecosystem models of vastly different sizes: from the Earth itself ("Biosphere 1") down to experimental "ecospheres" no larger than a basketball. One key conceptual forefather of Biosphere 2 dealt on a much smaller scale than planetary ecology. He was microbiologist Clair Folsome. Folsome participated in conferences at Biosphere 2 and advised the builders until his early death in 1989. His own experimental worlds were small enough to hold in his hands, but they provided a powerful analogy to Biosphere 2. In 1968 Folsome had gone to a Hawaiian beach; scooped up mixtures of algae, sand, seawater, invisible microorganisms, and air; and sealed the little sea communities in two-liter glass flasks. In his lab at the University of Hawai'i, he then watched as his miniature closed systems—"ecospheres," he called them—survived for decades with no inputs from the outside world. The biota in each ecosphere negotiated a balance, maintaining their own air and water chemistry within narrow ranges. The exact results were unpredictable, but life always found a way to live on. As Folsome excitedly told the Biosphere 2 team at their first conference, "It could just as easily be that if one were to take a small part of the world and to enclose it, even though it had appropriate energy inputs it could, for some reason or another, turn brown . . . In fact, it's hard to escape the conclusion that they almost always turn green."[4]

A faith that Biosphere 2 would similarly "turn green," and evolve to reach its own balance, propelled the builders forward. Biosphere 2 would simply be doing the same thing Biosphere 1 was already doing, they reminded each other. "A materially closed, energetically open system already exists; it's the Earth, and it's only a matter of technical detail to recreate such a system as a biosphere," Folsome told the 1984 Biospheres Conference. John Allen heartily agreed; Biosphere 2 would function much like an ecosphere in Folsome's lab, he wrote. "From Folsome's ecospheres to the Test Module would be a jump of some five orders of magnitude, a 'scaling-up' by a factor of 100,000. From the Test Module to Biosphere 2, the jump would be around three orders of magnitude."[5]

However casual Allen made it sound, there were still plenty of unknowns and assumptions involved in jumping across scales. For example, Biosphere 2, lacking a planet-sized atmosphere and ocean, would fundamentally differ in its basic chemical cycling patterns from Biosphere 1. No one could predict for sure how those differences would

Mark Nelson with Clair Folsome's ecospheres—which became the builders' tiny symbols that a closed biosphere would self-organize for life.

© PETER MENZEL / WWW.MENZELPHOTO.COM.

matter. As Taber MacCallum, a future biospherian, put it, "You take all those complexities away—a large ocean, upper atmospheric reactions; totally skew the air volume-to-land surface area, biomass-to-air volume, biomass-to-ocean, and every ratio you can think of . . . Do we have a pliable enough biota or ecosystem that you can really mess with it in a big way and still have it function?" As Tony Burgess put it, "I was told by one of my professors that I had the ultimate challenge in ecology: to make a functioning, evolving ecosystem from scratch."

The unknowns in that process were huge—as became clear in the emerging "green slime" debate over the new world's future. Once the doors were shut, would Biosphere 2 remain a pristine mixture of different wildernesses, or would it evolve into an undifferentiated mass of weeds and algae—or something in between? And what should (and could) the humans do about it? As the creators debated these questions, it became clear once again that Biosphere 2 was much more than just a science experiment.

Some consulting scientists enjoyed speculating on the "green slime" question. They looked forward to watching some species die and others survive in Biosphere 2, in order to gather ecological data. Conservation biologist Tom Lovejoy of the Smithsonian, an advisor to the project, actually remembered some scientists arguing at a meeting, "You'll learn *more* if it all turns to green slime." Famed systems ecologist Howard Odum publicly praised Biosphere 2 as an experiment, but he also publicly predicted that only 20 percent of its species would survive the first two-year enclosure period. Critics "predicted that within six months or so the whole biosphere would turn into a green slime. Not only would we lose the coral reef, but we would never keep the diversity of the rainforest, savanna, and desert," Gaie Alling recalled. "So we were in such unknown territory . . . This was so wildly out there, and people were not sure what was going to happen." Indeed, Biosphere 2 was being purposefully overpacked with species, in an acknowledgment that humans could not predict what would happen and that many creatures were bound to die off. Even the detail-obsessed biome designers were prepared for high numbers of extinctions. "All the people in charge of biomes agree that we're expecting a 50 percent extinction rate to occur in the two years—hopefully not humans," Ghillean Prance, the rainforest captain, announced at a 1986 workshop to a chorus of laughs from his colleagues.

Why would Prance and others take such an attitude even after going through such a painstaking process to select species for the ark? The answer lay in the biologists' faith that Biosphere 2 would act like one of Clair Folsome's little basketball-sized ecospheres; it would "self-organize" to find its own balance. Folsome had found, by analyzing seawater samples collected all over the world, that different species of microorganisms performed the same functions in each place. He advised the Biosphere 2 designers that they could collect air and water from anywhere they wanted in the world; no matter what, the microbiota would work out a way to maintain the system. Life forms would evolve their own equilibrium at every level.

In watching the "self-organization" process unfold, the Biosphere 2 ecologists believed, they would learn. The strategy, said Tony Burgess, was first, with faith in their own intelligence, to "spend a lot of time and money specifying initial conditions," planning out the wild, one plant at a time. Linda Leigh recalled watching Tony crouching in his field vest, meticulously mapping out wet and dry pockets of his desert to decide where to station each plant: "Each plant to him was a precious treasure." But after this careful planting, the team agreed that they would give in to faith in the Biosphere's own intelligence, said Burgess. "Listen to the system; see where it wants to go."[6] That would mean that out of the 3,800 carefully chosen species, some would survive, but vast numbers would not.

But here the scientific perspective occasionally ran up against other visions for the new world. John Allen acknowledged that many species might die off; after all, Biosphere 2 was an experiment. However, in comparison to some of the ecologists, he had less interest in sitting back as a spectator to miniextinctions. Allen was an engineer and a businessman, and he wanted Biosphere 2 to *work*. "Biosphere 2 was a metaphysical imperative for John Allen," recalled biospherian Taber MacCallum. Allen's 1985 book *Succeed* declared powerfully and in his own personal grammar that "the proper use of thought makes the possibility for *succeed* to become a way of life, given the possession of sufficient intention, attention, confidence, and intuition."[7] Throughout the Institute of Ecotechnics projects, Allen and his team had been working hard to develop their managerial abilities; now, at Biosphere 2, Allen seemed driven to make *succeed* a "way of life." As a result, Kathy Dyhr recalled, John might snap at any scientist who brought up doubts about green slime or species loss: "Well, is this guy on our side or not?!" When I asked John about the scientists who raised the "green slime" possibility, he told me he found them "lacking in ecosystem knowledge." However, in fact many ecologists urged that the evolution of Biosphere 2 would not truly be in the humans' control. As Tony Burgess said, "At every turn, Clair Folsome, the conceptual godfather . . . would not let us escape from the fact that ultimately the microbes were in control, they would always be in control." But, Tony recalled, not everyone at Biosphere 2—particularly those dreaming of building a productive and complete new world—appreciated that perspective: "Other people were disturbed by this notion that they would be basically trying to initiate some sustained dialogue with microbes, and it wasn't going to be easy. It was going to take a lot more time and expertise than anybody thought." John Allen was an avid fan of Folsome's microbe work; Folsome had proved that because of the microbes, in Biosphere 2 "there would be a teeming world of life adjusting the atmospheric composition to life's needs," John told me. But John also had his eyes on making history—fast. In a 1989 interview he declared the question that drove him: "How far, how deep, how complex can we go—and how quick?"[8] That approach did not always go over well with scientists. One hired scientific consultant from the Environmental Research Laboratory, David Stumpf, quit in anger after tangling with Allen. At a meeting, Stumpf announced his prediction that carbon dioxide levels in Biosphere 2 would soar to dangerous heights. Then, Stumpf recounted, John Allen "got up and started to rant and rave . . . He said, 'This has got to work. It's Gaia. The biosphere. The balance. It has to work.'"[9]

How would it "work"? Allen seemed hopeful that one species in Biosphere 2—the humans—would play a pivotal role. In interviews and speeches, he used a variety of concepts to describe humanity's ideal function in Biosphere 2. He spoke often of what renowned conservation biologist E. O. Wilson called the "naturalist trance," a state in which an able-minded observer would become absorbed into union with his or her environment. "In this particular state of attention, the scientific observer can begin to take in the totality of the life events occurring around him, and receive insights into its mechanisms and patterns," Allen wrote.[10] In such a trance, Allen believed, intelligent humans could intuitively guide Biosphere 2: "The trick is not so much in picking or choosing the species that will survive but in entering a naturalist trance in which the intellect is still working and perceives a pattern," he told a reporter. "It sounds mystical, but it is not. It is partly science, partly art."[11] John refined his concept of the trance with words from the animal behavior scientist Konrad Lorenz, who told him, "Find your own way to attain the naturalist trance, stay in it until you see something new, then record that observation. Classify and meditate on the data until patterns emerge." Allen resolved to take the trance to a much grander scale: "I wanted to observe the biosphere the way Lorenz observed geese, Wilson observed ants, and Goodall observed chimpanzees."[12]

At other times, John Allen used more engineering-oriented metaphors to describe Biosphere 2 in physical and mathematical terms. He frequently talked and wrote about uncovering "the laws of biospherics." Others recalled how Allen also spoke of discovering a "biospheric number" that would tell the state of an entire biosphere. Allen penned numerous scientific papers aiming to formulate the "laws" of biospheres, modeled on the laws of thermodynamics. In one paper he identified those "laws of biospherics":

> 1) Radiant and perhaps other forms of energies passing through a biospheric system increase the free energy in the system during the passage of time.
> 2) The biospheric system firstly uses part of the energy momentum to increase its potential to extract free energy out of the incoming energy flux converting ever more inorganic matter into organic matter (the "pressure of life" as Darwin called it); and secondly, converts the organic matter into an ever more highly differentiating array of molecules and metabolic systems capable of storing yet more free energy and extracting free energy more efficiently from the source to sink energy flux (evolution by natural selection).[13]

Still an unspoken question lingered: how could humans use biospheric "laws" to inform management? At one point, John Allen asked Tony Burgess to think up a formula that would spit out a precise number telling the managers when to "rain" on the Biosphere 2 desert. Burgess was confused; he liked to think about complex interactions of his desert creatures and their environment, but not in terms of single numbers. As a naturalist, Tony tended toward different metaphors; in contrast to the "naturalist trance," he described the Biosphere's human residents more humbly as thinking animals, just one part of the food chain. At one planning meeting in 1986, he likened the biospherians' role in Biosphere 2 to that of starfish in tide pools on the Washington coast:

"As soon as you remove the starfish from the community, the mussels overgrow everything else. But the starfish, by sampling whatever organism happens to be the most common, keeps any one organism from overcoming all the others. And in our case, the biospherians will be the keystone predators." Tony also liked the metaphor of "Earth Jazz"—an improvisational, constant conversation in which "you play four beats, nature plays four beats back."[14]

Perhaps the swirling of different metaphors was an indication of a broader American cultural confusion about the proper relationship between humans and nature. The Biosphere-builders were trying to work out humanity's correct role in nature. Could they really initiate and guide the evolution of an entire biosphere? Many myths that the Theater of All Possibilities played with on stage featured humbling morals about the limits of human power and the dangers of dreaming too greedily. Before building Biosphere 2, the actors toured performing John Allen's wild adaptation of *Faust*, the story of a man who seeks knowledge and magical power but has to sell his soul to the devil to get it. The troupe's adaptation of the epic of Gilgamesh played on similar themes. Gilgamesh, the hero, sets off in search of eternal life and meets one man who has achieved it—but Gilgamesh himself in the end is mortal. "Show me a man who can fly to Heaven!" he cries in the Biosphere-builders' play:

> Only the gods live forever
> Companions to the glorious Sun,
> But the days of us men are numbered,
> Our projects but gusts of wind.[15]

The actors and directors were consciously playing with parables. *Gilgamesh* fundamentally addressed "the seminal myth about the 'conquest of nature' that has driven Western culture's actions within the biosphere," said theater director Kathelin. Meanwhile, *Faust* was about "the exercise of power and knowledge for its own sake, without consideration of consequences," she said. In each of those scripts, published by the builders' own Synergetic Press in 1984, humans got smashed when they tried to become too much like gods. No one could say what would hold true for Biosphere 2.

Setting fiercely ambitious schedules, John Allen and his coleader, CEO Margret Augustine, showed their resolve to build and succeed quickly. They drafted complex diagrams of all the elements that needed to fall into place in a certain order in order to close Biosphere 2 on schedule. The diagrams became known as the "Critical Path."

But the critical path of Biosphere 2 kept having to be revised. Originally, in 1985, conference delegates had agreed that Biosphere 2 could be finished and closed for test runs by 1988 and that the biospherians would be inside by the end of 1989. As months and years passed, and as the complexities of the project multiplied, there would not be time for a total test closure; the biospherians would try out some simulations, and then once construction was done, get inside, lock the doors, and go for it. By 1988 the date for full biospherian closure was estimated as September 1990; in October 1990 it was reset for six months away, March 1991; finally it was pushed to its actual date of September 26, 1991. As CEO and head of design, Margret Augustine rallied her teams to meet near-impossible deadlines. NASA scientists, who visited the project site, compared the atmosphere to high-pressure spacecraft constructions and launches that they had experienced. John Allen found inspiration in the example of NASA's Apollo program. Its director George Mueller, who visited Biosphere 2, taught Allen the mantra "All up at once"—rather than painstakingly testing each component, just test the key subsystems, then put them all together and get the whole thing up and running. "It was always hammered on that you have to go faster, faster, faster," said Bernd, comanager of construction. "I was asked to give the time schedule, I gave them a time schedule, then there was a big meeting, and I was almost fired over it because it was far too long, and it was cut in half. And then it ended up to be exactly what I predicted it would be in the end."

No one could say exactly what the reason for the rush was. In early speeches to his companions, John Allen had insisted that a speedy design-and-build process would be crucial to take advantage of the historic "window of opportunity," before anyone else (such as one of the Cold War government superpowers) beat them to space, or some other catastrophe befell the Earth. But to some, the rapid pace was more a product of Allen's own personality. "John just operated that way because he got bored otherwise," said Kathy Dyhr, who had worked and lived with him for years. "John is an extraordinarily intelligent guy. He can't stand to puddle around with minutiae and people hemming and hawing and 'Well, let's think about it and have a committee report in six weeks.'" She pounded her fist on the table, mouthing his impatience: "No, goddammit!"

Or perhaps the push to succeed had more to do with the fact that Biosphere 2 was not just a project about the science of biospherics. As Linda Leigh put it, "The core ideas didn't have to do just with the Biosphere. They had to do with basic cosmology, with an understanding of everything—not just Biosphere 2 and not just the Earth." The basic theory underlying the project—namely, the idea that any enclosed biosphere would organize itself into a balanced and diverse system—had both physical and metaphysical levels. It was about microscopic organisms; it was about humans and their plans; but ultimately it was also about power and meaning in the universe.

On the physical level, the scientific rationale for Biosphere 2 came not only from the examples of Clair Folsome's ecospheres, but also from the work of the great early-twentieth-century Russian geochemist Vladimir Vernadsky. Vernadsky was a patron saint of Biosphere 2; drawings of his white-bearded, studious face graced the office walls at the building site. He had been the first to describe and define the Earth's "biosphere" as a unified, cosmically powerful, geological force shaping the planet. Importantly for the Biosphere 2 visionaries, Vernadsky always used the language of *power*: "There is no force on the face of the Earth more powerful in its results than the totality of living organisms," he wrote in 1926.[16]

Vernadsky's description of the biosphere as a powerful force differed from the language of later scientists writing at the time of Biosphere 2. Around the same time as Biosphere 2's invention, maverick scientists James Lovelock and Lynn Margulis were propounding their controversial Gaia Hypothesis—an idea quite similar to Vernadsky's concept of the biosphere, but phrased in very different terms. In the 1970s, Lovelock, who studied the atmospheres of far-off planets, and Margulis, who studied bacterial populations under a microscope, joined forces to popularize the theory that they named after the Greek earth goddess, Gaia. Rejecting the individualistic, Darwinian view of evolution,

Lovelock and Margulis proposed that evolution was a cooperative process, driven by countless species coevolving into a dynamic, interdependent planetary system. Microbes, in their theory, played a starring role in cocreating and maintaining planetary atmospheric chemistry. In their controversial 1975 article, "The Atmosphere as the Circulatory System of the Biosphere—the Gaia Hypothesis," Margulis and Lovelock first proposed that all organisms on Earth as a whole acted like cells, working together in a single living body. Lovelock's widely selling books spread the concept, starting with his 1979 work, *Gaia: A Look at Life on Earth*. In 1988 the American Geophysical Union even held a conference to evaluate the controversial idea.

But the Gaia Hypothesis as a scientific theory fell short of giving meaning to human life in the universe. The scientists' "Gaia" was not a goddess; the word was only a metaphor for uncountable masses of microorganisms working together. In fact, the Gaia Hypothesis attributed more importance to microorganisms than to human beings. And as John Allen later told me, he did not want to build "some Jim Lovelock–Lynn Margulis paradise for microbes"; a biosphere seemed meaningless to him without human culture. Perhaps for this reason, Allen and his colleagues invited Margulis and Lovelock to make speeches at Biosphere 2, but held more reverence for their long-dead Russian predecessor Vladimir Vernadsky. Toward the end of his life, Vernadsky had turned philosophical, suggesting that mankind must guide the biosphere's power. In 1945, in a paper called "The Biosphere and the Noösphere," Vernadsky described how man now had the power to transform the environment of the entire planet. People could create new metals and even change the chemistry of the atmosphere and the oceans, he pointed out. The Earth's biosphere was reaching a new stage in its evolution—and it was time for humans to play their role more consciously: "For the first time man becomes a large-scale geological force. He can and must rebuild the province of his life by his work and thought, rebuild it radically in comparison with the past. Wider and wider possibilities open before him."[17]

The new stage in evolution, Vernadsky wrote, would be the "noösphere": a biosphere guided by human intelligence. The suggestion that humans must guide their biosphere, acting together as a wise "geological force," appealed to John Allen and his fellow synergists. As Mark

Nelson told me, Vernadsky was more appealing because he "puts the responsibility very squarely on humans' shoulders and minds as having the responsibility to learn to become wise, careful stewards, since we have such creative-destructive power."

As the Biosphere-builders admired heroes like Vernadsky and philosophized accordingly, they were grappling with a great metaphysical challenge of the modern scientific age: to put meaning back into the science of biology. More than a century earlier, Charles Darwin, starting by explaining the diversity of finches in the Galapagos Islands, had begun the long scientific process of unraveling a sense of human purpose in the universe. As science historian Daniel Dennett wrote,

> Darwin's idea had been born as an answer to questions in biology, but it threatened to leak out, offering answers—welcome or not—to questions in cosmology . . . And if mindless evolution could account for the breathtakingly clever artifacts of the biosphere, how could our own "real" minds be exempt from an evolutionary explanation? Darwin's idea thus threatened to spread *all the way up*, dissolving the illusion of our own authorship, our own divine spark of creativity and understanding.[18]

If mindless rules of evolution could explain everything, then what was the purpose of life on Earth? If humanity was just another smart monkey, what was the point of human existence? Darwin and his intellectual heirs had torn down old biblical creation myths without substituting anything in their place. Neo-Darwinism reigned again in the late twentieth century, popularized by scientists like Richard Dawkins, who suggested in his best-selling book titles that evolution was simply a result of *The Selfish Gene*, that any God driving it was nothing more than *The Blind Watchmaker*.

The Biosphere-builders seemed determined to inject a sense of purpose back into one of the biggest ideas explaining the world: evolution. By building biospheres, John Allen and Mark Nelson wrote, they would "transform *Homo sapiens* into a creative collaborator with biospheres, rather than a parasite weakening the host"; they must "assist the biosphere to evolve." If they succeeded, they believed, they would become "heroes of a new kind—heroes who are champions of life and explorers of space."[19] Bernd Zabel remembered the first time he heard John Allen explain the idea of Biosphere 2 in a lecture. John described the history of the solar system, pointing out that the sun would eventually die. From that he drew a logical conclusion: if life was to survive eternally, it would have to go beyond planet Earth, and even beyond its present solar system—and someone would have to help it get there. And so, John told his listeners in all earnestness, their purpose: by moving to space, to alter "the evolution of mankind."

Esoteric ideas, perhaps, but they could be playful as well, as in the Theater of All Possibilities production *Tin Can Man*. In the play, a group of characters freak out about the state of the Earth and determine to take charge of evolution; the actors conclude with a countdown as they launch themselves singing into space. Extraterrestrial travel, in the play, was the obvious next step—a conscious step—in evolution, as one character put it: "First we climbed out of the water into the swamp, then we climbed up the trees out of the mud, then we climbed down onto the land, and now, at last, we're taking off!"[20]

That general plot of reclaiming the process of evolution, like so many other mythic dramas underpinning Biosphere 2, had strong parallels elsewhere in society—indications that the Biosphere-builders' fervent philosophizing was their own offbeat version of a widespread search for meaning. Various mystical gurus, such as the French Jesuit priest Teilhard de Chardin and the Indian yogi Sri Aurobindo, had gained large international followings in the early twentieth century using similar philosophies. Each of those gurus preached his own theological system based on a spiritual definition of "evolution," promoting a faith that through conscious spiritual effort humans could literally evolve. One of the guiding lights from the early days of Synergia Ranch, Gurdjieff, had become popular through the same type of appeal. Gurdjieff taught, as his disciple P. D. Ouspensky wrote in *The Psychology of Man's Possible Evolution*, "that man as we know him *is not a completed being*; that nature develops him only up to a certain point and then leaves him, either to develop further, *by his own* efforts and devices, or to live and die such as he was born, or to degenerate and lose capacity for development."[21] Biosphere 2 fell squarely in line with that Gurdjieffian view of evolution: humanity must pick up the work of creation where God had left off.

And so, as though determined to equal the original God's feat of creating Biosphere 1 in six days, the builders of Biosphere 2 rushed forward. Life sped up. "Our team's standing joke for years had been that 'a day off every three months (at the two solstices and two equinoxes) never hurt anybody,'" John Allen wrote later; but at Biosphere 2, the joke became reality.[22] Even long-term Institute of Ecotechnics members began to feel the strain. "Everyone was just having to rush, work harder, and we'd always worked hard," Ben Epperson said. Epperson had come in and out of the synergia network for decades, and at Biosphere 2 was helping to raise the rainforest plants. "It wasn't a factory," he said. "But it was getting to the point where it was almost like a seven-day workweek. Any leisure holes were being filled up, and there weren't that many to start with." Throughout the construction process, the designers were racing so frantically to draw plans for the Biosphere that sometimes the construction crews had to sit on site waiting to hear what they were supposed to build. Recalled Whitey Feigum, a local construction worker, "You'd get one set of drawings, at ten o'clock you'd get another set of drawings, at noon you'd get another set."

Meetings got tenser. At each Project Review Committee meeting, participants went around the circle and listed their top worries. Records from the September 1989 Project Review Committee meeting, when final closure was supposed to be only one year away, suggested the number of balls in the air at that time: there were close to three hundred construction laborers on site, working under eleven subcontractors; they were busy raising the Biosphere's skeleton of concrete, stainless steel, and spaceframe; construction of the on-site power plant, the "Energy Center" was also under way; and already the farm team had planted the first crops in the Intensive Agriculture Biome, even while the construction crew was still working to finish the arched spaceframe roof above it. Cranes towered over the growing Biosphere, lifting huge metal supports and heavy panes of glass into the frame of arches and pyramids. Meanwhile, people from every part of the project were gearing up the Test Module for another highly publicized test enclosure; this time future biospherian Linda Leigh would spend a full twenty-one days inside while her colleagues watched and analyzed the data. And even while all this was going on, the builders were trying to forge scientific collaborations. They cosponsored a conference in Siberia in 1989, on closed ecological systems, with the Soviet Institute of Biophysics, the scientists who had developed and tested small manned enclosures in the Bios experiments. In the same year they put on another conference at the Biosphere 2 site, in collaboration with NASA, focused on commercial applications of closed ecosystem technologies.

The creatures of Biosphere 2, meanwhile, were still scattered and coming together in a massive coordination effort. A team flew to Venezuela and rode in wooden dugout canoes up muddy jungle rivers to harvest rainforest plants. Back in Arizona, workers loaded these plants into quarantine greenhouses, then moved them into another greenhouse until the Biosphere's soils were ready for them. The scientific consultants and Ecotechnics members collected desert plants in Baja California and seeds for savanna grasses in South America and Australia. Still, a few hundred plant species for the wilderness remained to be chosen; the designers sat in long meetings hammering out the final food webs with the help of a team of entomology consultants. Meanwhile, far to the east, mangroves for the Biosphere 2 marsh were being collected in Florida, dug out of the swampy Everglades in huge chunks of mud and roots and loaded into trucks for their cross-country journey. Far to the south, the ocean team was still securing permits to dive and collect corals in the Bahamas and off the Yucatan peninsula. The *Heraclitus* crew had planned to collect the corals sooner, but during Hurricane Hugo in September 1989, the ship sank to the bottom of a harbor in Puerto Rico; an emergency salvage team of Biosphere 2 staff had to drop their tasks and fly there to help rescue the ship. Months later, in an effort headed by Gaie, the crew of the resuscitated ship would finally put on their scuba gear off the Mexican coast, dive into the ocean, choose their corals one by one from the sea floor, pack the precious corals in plastic trays full of seawater, and load them into the back of a truck for a three-and-a-half-day journey over bumpy Mexican roads to Biosphere 2.

At the same time, the international spotlight on the project was heating up. Biosphere 2 popped up more and more on the evening news and in the *New York Times*, Associated Press, and newspapers all over the world. Contracts were in progress for books and films to publicize the project. Hundreds of curious visitors began showing up, in increasing numbers, to tour the new world's construction site.

As construction proceeded, the costs piled up. Original cost estimates were proving to be way off. Early newspaper reports had announced the project's predicted construction budget as $30 million, but by the time Biosphere 2 was complete, the bills would total around $150 million. The huge number of events to coordinate sometimes led to huge excesses. The mangroves for Biosphere 2 had to be harvested before late summer hurricane season hit Florida, but when the chunks of marshland arrived in Arizona after being trucked cross-country, their spot in Biosphere 2 was not yet ready, so a construction crew had to build the plants their own climate-controlled greenhouse, complete with a saltwater circulation system—only to tear it down months later when the marsh moved into Biosphere 2. When a convoy of milk trucks arrived carrying Biosphere 2's seawater from the Pacific Ocean, it turned out that the truck tanks still had milk residue in them—so the ocean was poured out to water the road, and the trucks were sent back to California to try again.

Despite the careful attention to species lists, it became apparent that there would be no final control over what made it into Biosphere 2 in the end. The rainforest plants were installed while the walls of the Biosphere were still going up around them. As a result, local insects freely flew and crawled in from the surrounding desert and made their homes. Before the glass was in place, local Arizona deer nosed their way in to munch on the baby rainforest.

As millions of dollars rolled off the checkbooks, the source of those millions, Ed Bass, was an elusive figure, rarely seen at the SunSpace Ranch campus. He spent much of the year back home in Texas or ranching at the Australian Ecotechnics station, and flew to Arizona only for occasional board meetings. If Bass was concerned about the amount of his money that his friends were spending on their new Biosphere, he did not show it. "Since I keep involved on a certain level with a lot of different projects, and come and go and so forth, I've learned the most valuable thing I can do for the management is to keep my hands off from day-to-day decisions," he said in a rare interview at Biosphere 2, reasoning, "If you're not getting into something day-to-day everyday, I could be a real detriment to the management."[23] He put his trust in his long-time synergetic colleagues. When a *New York Times* reporter asked him if Biosphere 2's builders belonged to a "survivalist cult built around the magnetic personality of John Allen," Ed would only speak kindly about his mentor and friend: "I've never found anything to contradict the notion that he is a fine human being. He's an exceptional thinker, an exceptional intellect. He can take a lot of seemingly unrelated phenomena, disparate data, and put it together in meaningful patterns."[24]

"It was such a complex project, and the concept was so complex, that I don't even know who could have managed it," Silke "Safari" Schneider, one of the biospherian candidates, reflected. Biosphere 2 was testing not just an ecological question but a social question as well: what human organizational structure could possibly manage a complex ecological world? Many ecological thinkers have struggled to answer that question. Scientist C. S. Holling, for example, has suggested that human social structures should mimic ecological structures: Because ecological systems are dynamic, flexible, and decentralized, successful human organizations must manage them in a dynamic, flexible, and decentralized way.[25] But to the contrary, many participants in Biosphere 2 would be shaking their heads and asking themselves for years, Why, then, did the exact opposite of that idealized pattern appear in Biosphere 2's management? As Mark Nelson pointed out to me in defense of the project, every major high-tech corporation has a CEO and a hierarchical power structure; Biosphere 2 was no exception. But why was it that the more complex and expansive the Biosphere 2 project became, the more centralized became its management?

Little struggles over the ecosystems started to ignite into control battles—especially around that question of whether Biosphere 2 would "succeed." As time went on, anyone who speculated too loudly on possible failings of the new world risked being subjected to attacks of scorn and anger—usually from John Allen or Margret Augustine. The silencing of naysayers went beyond the "green slime" debate. Bob Scarborough, the soil scientist in charge of concocting Biosphere 2's dirt from quarry sites all over Arizona, had one such story. His bosses, John and Margret, instructed him that they wanted the biospherians to enjoy "a nice, soft Caribbean-type sandy beach" at the shallow end of the long, rectangular Biosphere 2 ocean. However, they also wanted a wave machine at the opposite end of the ocean to create enough turbulence to stimulate the corals, which liked to grow in wavy open seas. Bob argued that the waves would erode his beach, washing the sand away into the ocean; after all, the nature of beaches was to erode and change. "That soft white sand is going to get swept out into the bottom of the lagoon . . . and I don't care if you fill the whole beach up with aragonite sand, you're not going to have a beach," the soil scientist told his bosses. But, he said, "they didn't like that at all" and went into closed meetings without him. Soft sand was ordered from a dredging company, dug up from the ocean floor off the Bahamas, shipped on a barge to Houston, and then trucked overland to Biosphere 2. The construction team, following orders to install the beach, then "built it exactly the way it was in the plans, turned on the wave machine, and within two weeks the beach was totally gone," Scarborough recalled, laughing. A team had to work around the clock in the ocean to build a rock lagoon wall to contain the slipping beach. It was a potential emergency because the corals were already on their way up the road from Mexico; they needed to go straight into their ocean home, and if they were suffocated by sand washing off the beach, they could die. Despite the rock wall, the beach would continue to have erosion problems in the future.

In another example, Linda Leigh recalled what happened at one planning meeting when she voiced the worry that *Pythium*, a water mold, might infect crop roots inside Biosphere 2. "John Allen got furious. He said that I was becoming negative, that that would never happen," Linda said. He told her to scream out "'*Pythium, Pythium*,' as though it's something that has taken hold of your psyche and won't let go," she recalled. In the end, unfortunately, Linda was proven right: "After two years, *Pythium* was a big problem in the Biosphere. We didn't proceed to anticipate that it would be a big problem and deal with it." She sighed. "Stuff like that is just kind of crazy-making, and I don't know why those things happened, other than the saying that John would say: 'It's building being, it's letting you see yourself in a different way.'"

John Allen did make inspirational speeches about bringing all the life sciences together under one roof in Biosphere 2. In a 1986 meeting with the scientific designers, he rallied his colleagues: "The biospheric project is going to involve every philosophic, political, scientific, artistic profession on the planet . . . On the normal existential level, to me these groups are deeply divided, so deeply that if the ordinary ideological, theological, professional rap was used, there would be no hope for the united effort."

However, as much as Allen spoke of synergy—on both human and ecological levels—he also wanted to be the director. As Tony Burgess recalled, John and Margret "would bring all of us together in the workshops, and we would be communicating freely, and they would have a representative recording everything in each session." But, Tony said, after a time it became clear to the scientists that "the *project* was to be interdisciplinary, but the *synthesis* was to be restricted to inner management. All these consultants would be brought in and asked for expertise, but *we* would not do the synthesis of the expertise." Furthermore, Tony said, "I seemed to notice a pattern that when some system or group of people really started to get their act together and function, and it looked like they were really starting to come together as a group, John or Margret or both would step in, and there would be a crisis or a change or a switch in personnel, or something that centralized attention and control back to the top management."

That same pattern began to appear in information flows within the project. All along, the project leaders had maintained that Biosphere 2 could turn a profit as a corporate science venture. Now that became a rationale for extreme caution about sharing information with the outside world. The management hoped that they could patent inventions from Biosphere 2—perhaps the air-recycling technologies, or even the design of whole biospheric systems for space—and therefore they told their employees that knowledge about the project was "proprietary." The problem was that no one knew just what they would patent—and so no one knew just what was supposed to be kept secret. "There was this whole strange dynamic of what information do you give people . . . and what information do we need to hold onto because we're going to turn it into something," said future biospherian Jane Poynter, who was working on developing the Biosphere's agricultural systems. "So it led to this strange dynamic of trying to hide everything."

That guarded attitude quickly chilled relations with many in the scientific community. Early on, NASA scientists had kept a curious eye on Biosphere 2 because of the government's own interest in space colonies. Biosphere 2 representatives toured NASA facilities and talked to their directors; NASA cosponsored a conference at Biosphere 2 in 1989. But before long, no one from NASA would officially or publicly endorse the Biosphere project at all. The trouble began when NASA engineers, building their own sealed ecosystem project, designed an air chamber that resembled Biosphere 2's expandable "lungs." The domed metal lungs were the invention of the brilliant William Dempster, a long-time synergia member and Biosphere 2's head of engineering. When Margret Augustine learned of NASA's imitation of the lungs, she became furious and threatened to sue the government for infringing on Biosphere 2's intellectual property. Still later, a separate group of NASA scientists from Johnson Space Center wanted to do psychological research on the biospherians. Once they got out into space, NASA astronauts had often developed psychological problems, as individuals and as groups—problems that the government carefully kept out of the public eye. The space agency's psychologists now hoped to use Biosphere 2 to study the human dynamics of a small-group expedition close up. Future biospherian Taber MacCallum had set up the study with NASA, and the government promised to supply full funding. But when Taber presented the project to John Allen and Margret Augustine, their response was, "We don't want a whole bunch of shrinks looking over our shoulders," Taber said, and they turned the study down. That, in Taber's words, was "the last scoop of soil on top of the already nailed coffin" for the relationship between NASA and Biosphere 2. He heard from friends at NASA that "a memorandum went through all of NASA that all communication with the project had to cease."

On one level, the Biosphere project operated as a high-tech, disciplined business venture. John Allen described to me repeatedly the way he had set up a joint venture between Ed Bass ("Decisions Investment") and the small group of project directors ("Decisions Team"), and how they had contractually engaged Biospheric Design (headed by Margret Augustine) to design Biosphere 2, and the Institute of Ecotechnics (headed by Mark Nelson) to maintain contacts with outside scientists. It was a complicated, formal structure. In a letter to me, Allen described it as "an entrepreneurial organization which happens to be run by a management that uses that enterprise to finance its love of working in theater and the arts and finance, creating a new scientific line of thought called ecotechnics which concentrates in uniting the technospheric with biospheric science."[26]

Yet not everyone saw it that way. Linda Leigh, one of the most reflective biospherians, chose to explain to me the social network of Biosphere 2 in a different form. Speaking slowly and carefully, she was trying to figure out what had happened to the collection of people who had felt like such a close team, even a family. Because she had since been training in systems ecology, Linda was trying to diagram the social hierarchy of Biosphere 2. But in the end the diagram she came up with was instead a series of concentric circles, in which some people were "in" and others were far outside.

In the outermost circle were the peripheral contractors hired as extra hands to build Biosphere 2 or carry out other tasks. They knew they were on the outside. "The very large Biosphere 2 construction crew was marked by extremely high morale," recalled Mark Nelson, "precisely because they knew how extraordinary what they were being asked to do was—and how innovative and potentially historic the project was." Yet one construction worker, Rod Carender, felt that his bosses "always put on this facade that they were better than anybody." The construction workers were not the only ones who found themselves outside of the full "synergy." "Johnny used to say, '99 percent of the world is asleep, and we're some of the few people that are awake,' and it showed in their dealings with people," reflected Ren Hinks, a visual artist who was brought in to film the closure of Biosphere 2. When Ren wanted to find out more about the philosophy of the project so he would know what his bosses wanted him to produce, "The attitude was 'I'm working on a higher level, don't bother me,'" he said; he felt like "a peon."

On the next level inside, in Linda's diagram of circles, stood the scientists who designed Biosphere 2. "They just did their jobs," she explained. Numerous scientists consulted on the project as well, some paid and some as interested volunteers, experts on everything from space engineering to insect biology; many were "friends, colleagues, and inspirations in dealing with Biosphere 2 on a nonsalaried, voluntary, and continuously intellectually creative way," as John Allen put it. These scientists understood the main ecological goals of the project and participated fully on that level. However, they only vaguely knew what happened on site when they flew back to their own institutions or went home to their own families at night. Peter Warshall, the savanna design captain, felt frustrated. As a consultant hired for his biological expertise, he was reporting to longtime synergia members who were farther into the inner circle, even though they knew far less science than he did: "They had perhaps a high school level of biology . . . On the other hand, since we were consultants, *we* were supposed to listen to *them.* So it put us in that absolutely impossible situation of being in a sense their teachers and in a sense their servants."

There was a reason Linda drew her diagram of the Biosphere's social structure as circles, she explained: people at each level could glimpse what was happening at the next level closer to the inner circle, but could never enter completely inside. Desert scientist Tony Burgess described how he once got to dine with the Ecotechnics team when he first worked on their desert garden for the nightclub in Fort Worth in the early 1980s. As everyone gathered around the dinner table, John Allen began making guttural noises, and others started chanting. "It was like everybody had a personal ritual they were performing before the meal, and Margret was still trying to engage me in some kind of conversation because it was clear I was not comfortable and wondered what was going on," Tony recalled. After dinner, the actors put on a skit about a botanist in his honor. But he was rarely invited back into the inner circle.

Linda Leigh herself, and others who lived out at the site in the desert, found themselves at the next inner level of the project's social structure. Leaving her career and life in Tucson, "I moved out there and hardly ever talked to my friends in town," Linda recalled. These were the people who lived, ate, slept, and breathed Biosphere 2; many dreamed of becoming biospherians. They went by nicknames—Sierra, Gaie, Laser, Harlequin, Safari, Firefly, Flash. They participated in theater until life got too busy and the Biosphere itself became the theater production. With Margret Augustine as director, in a ranch house built on a hilltop across from the rising Biosphere, they put on *The Threepenny Opera, Deconstruction of the Countdown* (a montage work adapted from the writings of William Burroughs), and *Prometheus Bound,* featuring Augustine herself as a silver-painted, glowing Prometheus. Even amid the rush of construction, the builders used theater exercises to work on their emotional responses and try to become more effective people. When the actors drove into Tucson on Saturdays for a night on the town, Linda recalled feeling as though they were making anthropological observations on the society they had left behind—even though they had not yet entered Biosphere 2.

Within this tight-knit team sat a still smaller and tighter cluster of people, the core group: a handful of long-time synergia and Ecotechnics friends who had worked and acted together around the world, who knew each other intimately.

Resisting this notion of an inner circle, John Allen insisted on a more official corporate way of portraying the picture, writing to me in a letter,

> All overall design, building, and operating decisions at Biosphere 2 had to be made or approved by (1) myself, first as Executive Chairman from 1984–86, and then as Head of Research, Development, and Engineering for SBV and as Ecoscaper for BD [Biospheric Design] and Margret on all those decisions as Chief Executive Officer of SBV and as CEO and Chief Architect of BD, (2) by Margret, myself, Phil Hawes as co-architect of Bio 2, and William Dempster Director of Systems Engineering on all architectural decisions at that level, and (3) by Margret, Dempster, and myself as Biosphere Design Executives responsible for all engineering-technical decisions. (4) All contracts were signed and enforced by Augustine who had to get them approved by the Finance Committee of SBV (Bass, Harding, and her). The buck on anything operational stopped with me and Margret.[27]

On paper that was all true. But on Linda's paper, at the center of a diagram of circles, sat the name of one man: John Allen. "There were times when it was really tempting to cross that line," Linda reflected—to push further into the inner circle where the decisions were made—"and other times when I just wanted to go way, way out."

Linda and so many others told the same bitterly ironic story: just as Biosphere 2 neared completion—the ultimate construction of the dream—the old atmosphere of excitement and hope gave way. John Allen had always been known for his verbal outbursts, which he might unleash at any unsuspecting group member. For years his friends had accepted this as part of their process of developing themselves. "It was really one of the philosophies," said Bernd Zabel, one of the Biosphere's construction managers. "You don't have to tell how great you are, but in order to really keep you on track you need negative feedback . . . Everybody made mistakes, and the mistakes were actually highly discussed, not to punish people but to point out." But now the negative feedback came in frequent and stronger doses. In the words of future biospherian Jane Poynter, "Things began to get really weird around 1990."

Linda described the changed atmosphere at nightly dinner meetings: "Sometimes it would just end in John Allen yelling at everybody. And I say that because it happened so much; you'd just get lambasted. You would be told that you were a stupid idiot . . . It got more and more that way toward the closing of the Biosphere, so there was actually a kind of fear and dread going into some of these sessions."

John Allen and Margret Augustine held a strong influence on fourteen of their closest associates, the biospherian candidates, because everyone knew that the two of them would make the final decision over which eight people would become biospherians. "It was like a dark cloud following John Allen around," said construction worker Rod Carender. "Every time John Allen came in the room, people just got really nervous, especially the biospherians—they got *really* nervous. Personalities changed instantly . . . A lot of us resented him because he had this effect on these people. You know, you'd be talking, having a good time laughing and joking, and John would walk in, and they'd be somebody else. They would just hush up, not look up, not make eye contact with him."

Some of the key creators glorified the synergetic approach to building Biosphere 2. "There was a whole lot of dynamical, flexible and evolving informal management that occurred since the project was so complex that it demanded a whole array of creative scientists and engineers," recalled Mark Nelson. "Biosphere 2 was totally innovative in the way that its ecological scientists and engineers had to and did work together." Yet others told stories of bursts of command-and-control management. The construction team would build something according to Margret's blueprints, but she would say, "I don't like it, tear it down," recalled construction worker Whitey Feigum. "They would say they wanted a building, and they would agree on a set of drawings like a normal contractor," Rod Carender said. "And then we went through the process of the Margret walk-throughs . . . She'd walk in and say, 'Nah, that window has to be moved over a foot . . . that door, no, I like that door there, move it over.'" When the workers finished the shining white Mission Control office building across from Biosphere 2, it had rectangular windows and doors, just as the blueprints had ordered; upon seeing the building, however, Augustine decided that she actually wanted curved-top windows and doors to match the arched ceilings. She ordered all the frames to be ripped out and rebuilt. A high-roofed office and gift shop building on the campus had to be torn apart and rebuilt three times, Carender recalled. The plans first specified that the building have two floors, but when Augustine saw the finished building, she decided it should be only one story with a high ceiling, and had the builders rip out the second floor—then changed her mind and had them put the floor back again—then settled on a compromise, and asked that half the second floor be left in place and the other half cut away again.

Stories of excess during the Biosphere's construction were legion. Some construction workers reported going home with brand new carpets, oak tables, and building materials that had been tossed in dumpsters. The budget seemed to be "an open checkbook," said construction worker Jeff Boggie. Contracting out major design and construction projects to outside consultants also cost millions. The Biosphere's computer system, built for between $20 and $30 million, could have been done for $2 to $3 million, estimated Gary Hudman, who did much of the work designing the computer control systems. "It was just waste; 80 percent of this was going to Hewlett-Packard's pockets for no good reason," Hudman said.

Meanwhile, Phil Hawes, officially Augustine's coarchitect, quietly left the project as the building process drew to a close. He had designed some of the first adobe homes built by Synergia Ranch, had drawn the first sketches of Biosphere 2, and played a major role in the design. When I tracked him down long after his time at Biosphere 2, now in his sixties, he was still living in a ranch home in Oracle, Arizona, talking about building "global ecovillages" all over the world; sketches of castle-like ecobuildings lined the walls of his drafting studio. He wanted to talk about the future, something he had been doing his whole life. He did not want to talk about what happened between himself and Margret Augustine, or exactly why he left Biosphere 2, but he did have a little story. "I remember John one time said . . . 'If the challenge is too little, people get bored. If the challenge is too big, they'll split.' I said, 'Ah, that's exactly right, that's why I'm splitting,'" Phil said. "It was fascinating. I was having a great time. And then I realized I wasn't having any fun at all."

"There was a feeling, at least on my part, that we all participated in the design of Biosphere 2," recalled Bernd, who was busy coordinating among the many different contractors building the Biosphere. "You always had the feeling Biosphere 2 was ours. We believed in it. It was really high energy, and this is effective, contagious, if you get into a group which is really humming." Then, in the last year of construction, he said, something shifted. It went "from an avant-garde group to a seventeenth-century court, where it mattered who slept with whom, and how close they were to the king and queen, and then suddenly . . . some people were speared to be criticized, others got ridiculous promotions."

Bernd experienced that shift personally. When the final announcement of the biospherian crew members was announced at a press conference in September 1990, he was the smiling crew captain standing proudly in front of a team of eight matching red-suited explorers. But then, with only a few months left before their scheduled entry into the Biosphere, he and fellow biospherian Jane Poynter went before John Allen and Margret Augustine, nervous about presenting a frightening report: according to Jane's calculations, the Biosphere 2 agricultural area would not produce enough food to feed eight people well. The report would later be proven right—but Allen and Augustine promptly fired the two messengers of the bad news. Jane Poynter would soon be reinstated when no one could be found to take her place, but Bernd would have to watch from the audience when his friends finally entered Biosphere 2. Mark Nelson, the well-spoken chairman of the Institute of Ecotechnics, took over Bernd Zabel's slot. Bernd's old crew captain responsibilities were transferred to biospherian Sally Silverstone. An addendum to the project's official book *Biosphere 2: The Human Experiment,* mentioned that Bernd "had developed a medical problem which would preclude him from the first crew." Years later I asked biospherian Roy Walford, the doctor and eccentric elder statesman of the crew, how the eight biospherians were selected. "By what they could contribute," he said. "And how much they could be controlled."

Still, despite simmering conflicts, the builders remained devoted to creating their new world. The selected biospherians dove wholeheartedly into preparing for their mission. The field of candidates had begun with fourteen enthusiasts, many of whom had worked on the Ecotechnics projects for years together. Eight people was a magic number, the project's managers had decided, based on their own experience in setting up Ecotechnic projects and based on their readings on group dynamics, ranging from studies of Antarctic missions to Polynesian anthropology. The lucky eight who made the team were all adventurers—so much so that a later psychological study found that the biospherians actually scored better than NASA astronauts in terms of fitting an "adventurer" profile: high sociability, high levels of activity, low susceptibility to depression. The biospherian women's psychological profiles looked more male than typically female, much like those of female astronauts.[28]

All of the future biospherians took on major responsibilities in building some area of Biosphere 2. Each would have his or her own assigned role to play inside the Biosphere as well. Linda Leigh, who coordinated the assembly of the wilderness biomes, would still manage the terrestrial wilderness once she went inside. Her counterpart Gaie Alling, coordinator of the construction of marsh and sea, would continue to run the marine areas. The two other women inside Biosphere 2 would oversee the agricultural systems. Sally Silverstone would be in charge of the field crops and serve as official crew captain. She had lived in a commune back home in England, where she and friends grew their own food, and after college she had moved to the poorest part of India to volunteer in village development projects, but managing a farm on the scale of Biosphere 2 would be a new challenge. Jane Poynter, also British, had lived at the Ecotechnics station in Australia riding horses, and she worked on developing and testing the Biosphere's agricultural systems. She was also "Harlequin"—outspoken and enthusiastic, a singer and player in the Theater of All Possibilities. In Biosphere 2, her responsibility would be the domestic animals.

The men in Biosphere 2 would range in age from twenty-seven to sixty-seven when they first walked through the airlock door. The eldest was Roy Walford, the bald-headed doctor/artist/adventurer who had researched how to extend human life. He would take time off from his post at UCLA Medical School to serve as the crew's medical officer. The youngest male was Taber MacCallum, who had heard about the *Heraclitus* while he was traveling the world in his early twenties and joined the ship's crew. From the beginning, Taber said, he had wanted to be a biospherian: "It was very exciting to get in on something very early—it was only if you got in very early that you had a chance of being a crew member." Taber had worked on developing the Test Module and the analytical chemistry systems for Biosphere 2; he would be the resident chemist monitoring the Biosphere's vital signs in the state-of-the-art lab he had designed. Mark Van Thillo, better known as "Laser," had cosupervised the Biosphere 2 construction process as head of quality control. An electrical engineer by training, Laser would be in charge of maintaining the "technosphere"—all the machines keeping Biosphere 2 running. The final biospherian would be Mark Nelson, who had been writing, talking, and living the ideas of ecotechnics for the better part of two decades, from planting trees at Synergia Ranch to managing the grassland at Savannah Systems in Australia. Mark had originally been in charge of developing Biosphere 2's space and environmental applications. Inside, he would be director of communications, but also would have more earthly jobs, such as tending the "wastewater garden" that would recycle the biospherians' sewage through a series of plant-filled tanks.

Each biospherian was required to master his or her own specialty area and complete training in a long list of other areas, including naturalist observation, analytical chemistry, computer programs, farming, mechanical maintenance, public speaking and writing, and "chef-level cookery." In addition to this training, however, the team members were to learn to work together in any situation; life experience was to be their real school. "Traditional psychological testing really doesn't play a role in Biosphere 2," Roy Walford told a reporter. "A better model is to just put these people in normal work situations and have them work out conflicts the same way we all do every day."[29] According to that logic, each biospherian candidate had to be practiced in the rhythms of synergetic group living and theater; make it through a stint sailing the oceans, sharing tiny quarters on the *Heraclitus*; and survive another stint doing tough horseback riding in the heat at the Ecotechnics station in the Outback. "Between life on the heaving ocean and in the baked savannah, all biospherians underwent training in extremes. These were our rites of passage," wrote biospherian Jane Poynter.[30] "You were just thrown in as another cowboy," Roy Walford recalled. "Not a dude ranch, a regular working ranch. The cowboys would be aboriginals—really rough-drinking, hard-riding cowboys. . . . You had to know how to ride, or learn to ride, and drive stakes and string barbed wire across the landscape, and repair windmills, and herd cattle, and castrate cattle, and get along with the aboriginals and drive bulldozers." "That," he said, "was the psychological test."

And so the biospherians plunged forward toward the closure day. Linda Leigh recalled their resolve: "When it was getting towards the time when we were going to go inside, several of us said, 'Let's just stick this out until we go inside, and then we don't have to deal with any of the management from the outside . . . Let's just get inside; we know how to run this place, and we know what's best for it, and let's just get through this.'" Her friend Jane Poynter put it more bluntly, referring to John Allen: "He was scarily wacko, but we were so committed, we couldn't put the brakes on at that point."

The moment of take-off approached at last. September 25, 1991, was the last day before "closure." The biospherians were still dashing around bringing in decorations for their apartments and posing with forced smiles for final photos. Teams of workers scurried to sweep out the trash left over from construction and scaled the walls to scrub the agricultural biome windows one last time to maximize the light reaching the crops. Two invited Lakota Indian elders built a sweat lodge on the ridge behind the Biosphere and led the biospherians inside for a sacred ceremony. At night, fireworks burst into the sky over the completed Biosphere 2 in a huge celebration. Two thousand guests danced on the lawn in the shadow of the pyramids and gleaming white tower. Psychedelic godfather Timothy Leary and actor Woody Harrelson were among the celebrities mingling in the crowd.

The next morning of September 26 dawned sunny and clear. A crowd of thousands of spectators arrived early to watch the final rite of passage from the lawn under the curving white walls of the Human Habitat. True to the builders' long desire for a synergy of world traditions, in front of the crowd the biospherians received blessings and prayers from special guests: Lakota Indian elder Dan Old Elk in his feathered headdress, a traditionally dressed Huichol Indian woman from Mexico, and a Tibetan Buddhist monk in maroon robes. (Meanwhile, a construction cleanup crew was still inside Biosphere 2, frantically putting on finishing touches and hurrying out the back door.) The biospherians themselves stepped to the microphone and made short, carefully worded speeches about the new frontier. "I will not return from this expedition the same," said Sally. Taber spoke of "the great American spirit of free enterprise and innovation that has made this project really go." "Listen, there's a hell of a good universe next door," said Roy, grinning from ear to ear, his bald head glinting in the sun. "Let's go."

Camera shutters clicked and television helicopters swooped overhead, and the eight biospherians, in their matching navy blue jumpsuits, strode in a single-file line toward the door hatch of Biosphere 2. Their friends and fellow builders, who would remain outside for the two years, stood along the carpeted path to give last hugs. Last of all, at the door to Biosphere 2, grinning in brimmed hat, coat, and tie, embracing each biospherian in a final bear hug, stood John Allen. When all eight had stepped through the airlock, they pulled shut the big metal door and together swung its heavy latching handle down and into place.

The miniature world as envisaged by its designers, flanked by its two white "lungs."

SYNERGETIC PRESS.

Biosphere 2: White Human Habitat with tower, agriculture biomes with arched bays, Santa Catalina Mountains in background.

ROY L. WALFORD LIVING TRUST.

Tony Burgess leads a crew in planting the desert.

ROY L. WALFORD LIVING TRUST.

Biospherians in the Intensive Agriculture Biome.

ROY L. WALFORD LIVING TRUST.

Mark Nelson with "R2D2," measuring soil respiration.

ROY L. WALFORD LIVING TRUST.

Gaie in the ocean.

© PETER MENZEL / WWW.MENZELPHOTO.COM.

Linda in the Biosphere 2 rainforest.

ROY L. WALFORD LIVING TRUST.

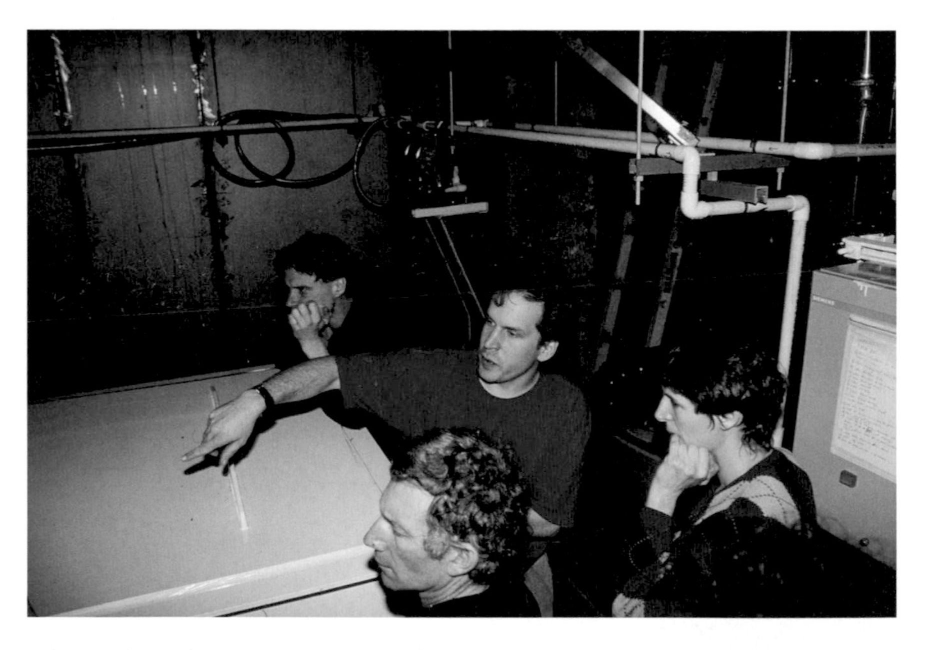

Taber explains the carbon dioxide scrubber machine to his fellow biospherians.

ROY L. WALFORD LIVING TRUST.

One of the crew's cherished feasts, on their beach.

ROY L. WALFORD LIVING TRUST.

John Allen (in white hat) looking in at the biospherians' visiting window with guests.

ROY L. WALFORD LIVING TRUST.

The short-lived Butt-Wheeled Wagon of Biosphere 2.

ROY L. WALFORD LIVING TRUST.

Roy and Jane in their music video "Ecological Thing."

ROY L. WALFORD LIVING TRUST.

Above:
Biospherians before their mission, from left to right: Bernd Zabel (replaced later by Mark Nelson), Sally Silverstone ("Sierra"), Abigail Alling ("Gaie"), Taber MacCallum, Mark Van Thillo ("Laser"), Linda Leigh, Jane Poynter, Roy Walford in front.

Biospherians at their "reentry ceremony" after emerging from two-year closure, from left to right: Gaie, Roy, Linda, Mark Nelson.

The marsh biome, during Columbia's management. A plastic curtain partitions the rainforest in the background.

PHOTO BY MARK GEIL.

Tourists in the Biosphere 2 desert, long after the biospherian missions.

PHOTO BY MARK GEIL.

ACT III

Biospherians at work in their farm, the Intensive Agriculture Biome.

ROY L. WALFORD LIVING TRUST.

PIONEERING

THE BIOSPHERIANS STOOD A LONG TIME JUST INSIDE THE AIRLOCK DOOR, looking out the window, waving at the spectators who had come to see them off. They stood only a few yards away from their friends and families, but already felt themselves far, far away. For despite all their differences of opinion, they clung to a common conviction: they were standing in a new world now.

"It was actually at that moment in which the engineer locked the airlock door that our glass world became a separate entity from the Earth," biospherians Gaie and Mark later wrote. "The Biosphere became a distinct material entity, physically closed off from the rest of the universe except for energy and information flows."[1] Taber, the youngest biospherian, who had long dreamed of space, imagined he was now stepping onto a space station: "You're making your own water, you're growing your own food, you're maintaining your own atmosphere, you're doing all the things that you need to do to be in space." Laser, who also dreamed of going to space one day, wrote, "I consider this more like a trip on a paradisiacal planet."[2]

The biospherians would not be physically going anywhere—in fact, they would be extremely confined in physical space—but that did not seem to matter as they imagined themselves as explorers. Sally, the crew captain, wrote in the *Biosphere 2 Newsletter*, "Like all pioneers, we are exploring new territory."[3] Linda, writing in her own journals, called herself "The Naturalist in a New World." Later the biospherians would talk via ham radio to a crew of scientists overwintering in a geodesic dome at the South Pole. They exchanged stories of their daily lives and envisioned themselves as kindred teams on far-out expeditions. Inside

Biosphere 2, the biospherians proudly displayed an expedition flag on loan from the Explorers Club, an exclusive association of professional explorers. The crew members' spacesuit-like wool jumpsuits, one set red and one set blue (which the biospherians would never actually wear inside the steamy wilderness of Biosphere 2), deliberately played on the drama of space travel. So did the official language of expedition used on the Biosphere 2 site: the biospherians were a "crew" with various "officers" and "captains," their two-year enclosure was dubbed "Mission One," and their bosses sat at "Mission Control."

As explorers, the crew hoped for more than new sights. Laser described to me the questions motivating him in Biosphere 2: "If all this works with me inside of it, or with eight people inside of it, it is going to give me experience that is going to give me information that I would really like to know about life . . . What do I need for me to be absolutely and totally content and happy? What would it take? Could I do it on this 3.2 acres?" His words echoed the hopes of many past men who had undertaken wilderness quests, epitomized, for example, in Henry David Thoreau's declaration of why he went off to live alone in the Walden woods: "I went to the woods because I wished to live deliberately, to front only the essential facts of life, and see if I could not learn what it had to teach, and not, when I came to die, discover that I had not lived."[4]

Such dreams of confronting life's essence were largely the fantasies of people raised in the comforts of civilization. There was a reason: pioneering, actually living on the land, is hard work. The first North American frontiersmen and women had to clear land, plant crops, build houses, and face summer droughts and winter storms. Likewise, the biospherians, despite romanticizing the pioneer lifestyle, in reality became eight extremely busy high-tech peasants. Each morning they got up early to gather around the long granite dining table, eat their porridge, and plan out their tasks for the day. After breakfast they labored at raking, digging, planting, and harvesting on their farm, the "Intensive Agriculture Biome." As the sun rose high through the arched glass ceiling, they tended to their eighteen square plots packed with different crops—potatoes, grains, beans, vegetables, even a rice paddy with fish swimming in it—stopping only to snack on a rationed handful of roasted peanuts. The rest of the day they took care of everything else: machines, wilderness plants, animals, data collection, and anything that needed attention—and something always needed attention. Each biospherian spearheaded crews in his or her own specialty area. Sally designated tasks for the farming crews and organized the food stores. Gaie supervised dives in the ocean to look after the corals. She got others to help her with the perpetual task of cleaning green slime out of the "algae scrubber" ocean-cleaning system in the cement basement under the concrete cliffs. Mark splashed around in another corner of the basement, under the agricultural biome, tending the "wastewater gardens." Roy ran medical experiments on his companions, often asking for samples of their bodily fluids to analyze in his well-equipped lab in the Human Habitat. Jane took care of the chickens, pigs, and goats in their pens downstairs from the biospherians' apartments, under the Habitat. It took a large effort to raise animals for a little milk, eggs, and meat: finding plant scraps for the animals to eat, feeding them, milking the goats, and on special occasions, slaughtering a pig and butchering it by hand.

The biospherians' labor was not glamorous. They passed long hours not just in their fields but in their concrete-walled basement, whether threshing grain or scraping muck off algae scrubbers. But through the dirt and sweat, the biospherians developed an enormous personal, emotional sensitivity to their world, like mothers caring for a demanding child. Each day Linda made her rounds through the terrestrial wilderness, giving each biome its health checkup. She waded through the dense rainforest, around the concrete "cloud forest" mountain, along the tiny snaking stream and pond, checking the soil moisture as she went and poking through the plants for wilting or brown leaves, which could indicate soil chemistry problems. From there it was a short but diverse walk to the other end of the wilderness: through the savanna's thick grasses, on cliffs hanging high above the ocean, and then down steps among boulders into the desert pyramid at the Biosphere's far south end. Walking through the desert and thornscrub, Linda peered around for green buds and flowers—signs of plants coming out of dormancy—to help her decide when to "rain" on them. From biome to biome she regularly made expeditions with an instrument affectionately named R2D2, a Plexiglas case housing monitoring instruments. Placing R2D2 on the earth, Linda could read how much carbon dioxide was coming out of the highly active soil, a constant concern in the Biosphere's closed atmosphere. Linda preferred to take each measurement herself to make sure they were accurate. Each day she also fed monkey chow and dried fruit to the galagos, Topaz, Opal, Oxide, and William Kim, the Biosphere's little tree-dwelling primates, who became like the biospherians' pets. She watched them as they learned to climb the spaceframe walls. As often as she could, Linda wrote down field notes, thoughts, and sensations in her journals. On her fortieth birthday, she got up early and watched the sunrise through the glass, back on Earth.

While Linda immersed herself in Biosphere 2's wilderness land, Gaie dove into the waters. She too felt an intimate connection to the ecosystems under her care. "The coral reef was my baby," she told me. Every day she dove in the little sea, swimming around to check the baby corals, make sure they hadn't been knocked over, weed out algae, scrub the underwater windows so that visiting tourists outside could see through the glass into the murky ocean floor. When the seasons changed, Gaie noticed that some of the corals looked bleached from the change in light levels, as they were used to the tropics, where they would receive about the same number of sunlight hours every day year-round; so she began a program of carefully "migrating" the corals with the seasons, keeping them in shallow waters where they would get more sunlight during the winter, and putting them back in shady spots for the too-long days of summer. While thinking and operating scientifically, Gaie was more than just a manager. With devotion, she nurtured her aquatic world from conception, through growth and construction, to daily life:

> It was never something that I could leave. So for at least six years, I literally never left the site. If I'd go into town, even for an evening, I'd have to make sure someone was there to respond to the coral reef, because if the waves stopped, if there was a burst in the savanna overhead irrigation and water flooded the ocean, if pumps went off and started spewing out saltwater into the basement of the savanna systems and contaminating the rest of the Biosphere . . . It was a twenty-four-hour, six- to eight-year journey of learning to live with a coral reef, from the nutrient controls, temperature controls, salinity controls, to the algae management.

Throughout her time in Biosphere 2, Gaie would summon other

biospherians to help "whenever there was a crisis in the ocean," recalled Sally—and between the risk of saltwater contaminating the rest of the Biosphere and the risk of the rest of the Biosphere contaminating the ocean, "there was always a crisis in the ocean." But Gaie still took pleasure in her sea. Parts were murky and full of algae, not quite like anything she wanted to recreate in nature, she said. "There were moments when you got down to the deeper ocean where you'd sort of look around and think, where am I, because there'd be a lot of algae, and not so much of the corals; it sort of changed its appearance, not exactly like a coral reef in the wild." But at other times, even as she swam around in her "ocean" not much bigger than an Olympic swimming pool, she could maintain for herself the sense that she was in a recreated paradise: "There were moments when you could be in the coral reef, and you would think, my God, I'm in the Caribbean, I'm in Mexico, and you would lose sight of the walls and you would be immersed in it. It would be just the most thrilling, incredible experience with the waves and everything, feeling that we really did it."

All day the biospherians paid attention to their miniworld's vital signs as though they were their own. They read out the state of the atmosphere at breakfast. As they made their rounds, checking plants and machines with their hands, eyes, and ears, they came to see themselves as sensitive human extensions of the computerized monitoring systems, said Linda. "We had the analytical instrumentation, but the people on the inside would be the instruments; our experience would be informing us." This sharpened ecological attention was born out of love, but also out of basic necessity. The biospherians felt their Biosphere was their responsibility, but they also knew their survival depended on it—which would become increasingly clear as time went on.

It was not just an intellectual or even emotional connection between people and ecosystems; the biospherians felt themselves bonded to their world with their very bodies. While her analytical mind focused on helping the corals grow, Gaie could also feel herself as an organism adjusting to Biosphere 2, she said. She felt like her body was physically undergoing a "molecular kind of change" after several months inside, as though her own organs and cells were adapting to Biosphere 2's unique chemical pathways: "I could feel a very different tempo, different quickness, different vibration." The feeling made sense, she reasoned, because "Biosphere 2 is operating on very different cycles than Earth . . . Actually adjusting to the matter, flowing that fast, which is much faster in Biosphere 2 than it is on Earth, would take an impact on one's body." The proportions of earth, biomass, water, and different gases inside Biosphere 2 were assembled in different ratios from those outside in Biosphere 1, so that liquids and gases cycled to very different rhythms; without the moderating effects of a large planetary atmosphere and ocean, the cycles were highly changeable and speedy. The biospherians had become part of that system, not just studying the test tube but eating its fresh vegetables, breathing its humid, carbon-dioxide-rich air, and drinking its recycled water; they literally were part of Biosphere 2. At the dinner table one night, Taber, the analytical chemist, joked about getting tired of having to eat the same atoms over and over. Working long days and eating slim harvests, the biospherians began to lose weight. As the pounds dropped off, Linda observed in her journal, "The Biosphere owns your molecules."

That blurring of boundaries between the personal and the scientific happened with each breath the biospherians drew. Every day they thought about their atmosphere—they had to, because carbon dioxide levels were on the rise. Linda recalled how from her first stay in the Test Module, she had become acutely sensitive to how each little action might impact the air around her. Just pulling roots out of the soil, her mind automatically thought in chemical equations: "If I would dig my sweet potatoes for the week, I would be disturbing the soil, which creates more carbon dioxide in the atmosphere. And I could tell that at the end of the day on my graph of the CO_2 that I had increased it by my digging. So, I would try to organize my harvesting based on when there was most light, or sometime during the day when it made sense in terms of carbon dioxide."[5]

Now, in Biosphere 2, that kind of thinking became automatic, out of necessity. Rushing to enter Biosphere 2 as soon as it could be finished, the crew had begun their mission just after the autumn equinox. From then on, the days would only grow shorter and shorter, toward the lowest light levels of the year. That meant that for the first few months, plants would be growing less and less, and sucking up progressively less carbon dioxide. Desperate to keep the plants producing as much new growth as possible before the winter, the biospherians hacked away at the dense branches along the rainforest's outer banana belt to stimulate the plants to grow back again.

The biospherians and their colleagues outside at Mission Control worried while CO_2 graphs on their monitors spiked upward. John Allen sent in messages by phone, video conference, and computer to suggest how the biospherians should manage their air. One day Allen declared a "carbon dioxide emergency," sending all the biospherians to plant more plants and install extra grow lamps all day. His campaign to control the CO_2 cycle had begun even before the biospherians' entry, with all sorts of schemes. Just before closure, he had even recommended removing all earthworms from Biosphere 2 because they breathed too much, Linda recalled. When Linda disagreed with what she saw as a "ridiculous idea," she said, "he fired me from being the scientific director inside," and Gaie became official scientific director. Still, now Linda stood in the desert across the window from her trusted friend Tony Burgess, communicating in written notes held up against the glass, to decide how she would water the desert to get its plants to suck up carbon dioxide in winter.

In such a small world, everything was inescapably connected to everything else. The connections often showed up in surprising ways. The source of the Biosphere's air problem lay buried in its soil. Biosphere 2's wilderness was planted in a mix of soils dug up from all over Arizona, stirred together in different recipes for different parts of the various biomes. The rainforest and agricultural soils contained ultrahigh doses of nutrients from a generous serving of "Wilson's pond soil"—a euphemistic name for manure-rich muck dredged from the bottom of cattle drinking ponds in desert pastures. The soil designers had purposefully made the Biosphere's soils unnaturally rich. Tropical rainforests naturally thrive on nutrient-poor soils in Biosphere 1; in the tropics, nutrients are stored in vegetation itself, not in soils, where they would easily be washed out by intense rains. In contrast, Biosphere 2's rainforest took root in rich soil packed with organic matter. But this strange design strategy had its own rationale: the builders believed that they were sealing Biosphere 2 for one hundred years, and during that time, baby trees had to grow tall and many cycles of agricultural crops had to grow over and over on the same ground. The carbon molecules to build the maturing plants had to come from somewhere, so they needed to be stored in Biosphere 2's soils from the beginning.

However, once the doors to the Biosphere were sealed, soil microbes feasted on the extra organic matter in the ground, and multiplied—and breathed. As the ecosphere-maker Clair Folsome had predicted, Biosphere 2 would not die; it would "turn green" with life—but also as he had predicted, the microscopic organisms, not the humans, would guide that process. The happy soil bacteria, as they ate and breathed, sucked oxygen out of the atmosphere and spewed out carbon dioxide, influencing the chemistry of the entire Biosphere.

Making things worse, the crew's first winter under glass turned out to be an unusually cloudy El Niño season in southern Arizona. The clouds outside blocked even more sun than expected, slowing plant growth inside. As a result, CO_2 levels inside Biosphere 2 at times climbed as high as 4,500 parts per million—more than ten times higher than in Biosphere 1's atmosphere outside. As the CO_2 accumulated, oxygen gradually dwindled. The biospherians were left breathing thinner and thinner air, as though they were slowly ascending a mountain over the course of several months. The risks extended beyond the biospherians' own lungs to the health of other creatures as well. For example, the ocean began absorbing the excess CO_2, forming carbonic acid in the water. If the ocean became too acidic, the water itself would gnaw away at the corals' skeletons, stunting their growth or even killing them.

As they nervously watched clouds cover the sun, the biospherians tried to use all available labor, space, and technology to change the equations for photosynthesis and respiration to their advantage. They constantly monitored soil moisture and temperature; if these increased too much, so would soil microbes' respiration. But the key variable was light; light would determine how fast and how well plants could grow, and how much carbon dioxide they would drink up in the process. Biosphere 2 sat in sunny southern Arizona, but its thick glass panes and the steel beams of the spaceframe blocked out half of the incoming sunlight before it could even reach the plants inside. To take advantage of all remaining available "sunfall," the biospherians covered the ledges and balcony above the agricultural area with potted plants. Mission Control sent in more lamps from outside to grow more plants in the basement. Each plant's biomass represented carbon atoms that would no longer be in the atmosphere. To keep the wilderness biomes growing vigorously, the biospherians threw themselves into pruning. Hacking through the savanna with sickles, machetes, and pruning shears, they repeatedly cut back tall grasses to encourage new growth. They dragged the dead plant matter down into one of the huge domed lungs to dry, so that it would not decompose and release carbon back into the atmosphere; then they baled the cuttings for storage in the basement. When the doors to Biosphere 2 were finally opened two years later, the cleanup crews would discover hallways lined with huge bales of dried plant matter. Alongside the bales of biomass would sit hundreds of barrels of white powder produced by a "scrubber" machine that removed some CO_2 from the atmosphere through a series of chemical reactions. And so the biospherians, despite their pride in how Biosphere 2 recycled everything, ended up with a huge basement full of components of their air that they wanted to keep out of the cycle.

Managing the flow of water inside Biosphere 2 was as theoretically simple, and as practically complicated, as managing its air. With no water coming in or out, Biosphere 2's water could be recycled indefinitely, the same water raining on the crops and rainforest over and over, running through the savanna and rainforest streams, flushing through the biospherians' toilets, and being purified by plants in the "wastewater garden" tanks in the basement. At least that was the theory. But all those different elements of Biosphere 2, squashed together into a single compressed water cycle, required different qualities of water. The humans, top priority, got to drink the cleanest water, which was collected as it condensed out of the atmosphere on coils inside the air-handling machines. However, some of the ecosystems were sensitive to water quality as well. As the endlessly-looped savanna and rainforest streams flowed over and over through their assigned channels from cascades to collection ponds, water evaporated from them, causing the streams to become increasingly saline over time. Meanwhile, in the ocean, corals proved nearly as fragile as the humans in their need for clear water. The marsh water, flowing into the miniature sea, was thick with organic compounds, tannins, and acids, which tinted the ocean yellow. The corals grew slowly, homesick for the clear waters of the tropics. The algae scrubber system, designed to prevent this problem, proved inadequate. In the dimly lit scrubber room under the oceanside cliffs, ocean water swished through trays filled with algae, which were supposed to eat excess nutrients out of the water—but in fact the algae created organic runoff of their own. After a few months Gaie and her ocean consultants outside decided they would have to segregate the ocean water from the marsh, but the ocean had already turned murky with excess nutrients. Finally Gaie and Laser devised mechanical "protein skimmers" out of PVC tubes and spare parts. As seawater bubbled through the skimmers, they would suck up unwanted organic compounds.

The supercompressed water cycle was wreaking havoc on Biosphere 2's land-based wilderness as well. The high CO_2 in the atmosphere meant that the wilderness "rain" was loaded with carbonic acid. The acid rain ate away at minerals in the rainforest's granitic soils at extreme rates—two hundred years' worth of rock weathering took place in two years in Biosphere 2. As the dissolved rock elements were washed down through the soil, they recrystallized to accumulate in a thin layer of salts that made parts of the soil impermeable to water. Runoff from the rainforest's soils, loaded with dissolved salts, drained down through the soil to sumps, then into pipes and pumps that carried it back into a giant round pool in the South Lung. As a result of this inflow, the lung reservoir—the Biosphere's main reservoir of fresh water—became saltier over time. The purest water, collected by sucking water vapor out of the atmosphere on the air-handling machines, was never as much as the biospherians wanted. Though they hoped to model a harmonious ecofuture, instead the biospherians found themselves modeling a more dystopian future—fighting wars over water. They quarreled over where to use the scarce supply of fresh water, which still had to be mixed with some salty water to produce rain. Linda coveted the pure rain to keep the wilderness healthy; Sally insisted on using it to water the food crops. And meanwhile, all prayed for their plants to grow. When Linda turned on the overhead sprinklers to awaken the desert for its winter growth season, all the biospherians dropped their work and came to stand reveling in the "rain" until they were soaking wet; tourists stood outside gawking, only dimly comprehending the serious hope for survival behind the lighthearted moment.

While all that chemical complexity looped around in the biospherians' minds, another simpler equation was growling in their stomachs: food, food, food. The biospherians were hungry. They had built a sophisticated wilderness to answer lofty questions; now they spent their days learning what billions of the world's poor could have taught them already: what it's like to be a hungry farmer. The chemical equation was an obvious one, said Sally, who was in charge of the agricultural fields: "There was only so much light coming in, and you can only convert so much of that to edible calories." Linda offered an even simpler equation: when the biospherians first entered their new home, "We just plain were not good farmers." Hunger would provide the motivation for them to learn quickly.

Growing enough food to nourish eight hard-working people on only half an acre would have been hard even in Biosphere 1. It became harder thanks to the low light levels and other idiosyncrasies of the Biosphere 2 farm. Under the enclosure's unique conditions, pests virtually unknown in Biosphere 1 thrived in Biosphere 2's fields, and pesticides were of course off-limits. Uninvited broad mites exploded into an infestation. This small mite was known in Biosphere 1 only in mountainous tea plantations and in greenhouses, because it thrived under the rare combination of low light and high humidity levels—and so the mite found paradise in Biosphere 2, where it devoured the biospherians' entire white potato crop. The only way that head farmer Sally could figure out to control the mite was to reduce humidity using the powerful air-handling machines above the farm, but that consumed huge amounts of energy. In desperation the biospherians even went through the fields trying to blow bugs off of the plants with hairdryers. Molds and mildews also became a problem in the humid atmosphere, and constant recirculation of the same air made the situation even worse, spreading spores that attacked the beans and the squash. The only bean hardy enough to produce reliable harvests in Biosphere 2 turned out to be the "lab-lab" bean, otherwise known by biospherians as the "blah blah" bean for its taste.

As if being farmers was not enough work already, the biospherians had to become pollinators too. Without wind or the right insects to pollinate the crops, many species would not produce food. The biospherians waded through their muddy rice paddy, whacking the rice stalks with sticks to make them release pollen. They bent over each tomato plant with an electronic buzzer in order to make the flowers jiggle—something bees or wind would do in Biosphere 1—to convince the plants to produce pollen. And under the low light levels, they could only watch hungrily as their plants grew exceedingly slowly. Altogether, the biospherians were learning the hard way the truths of ecological interdependence. As several remarked, none of them had so thoroughly appreciated their interconnection with global atmospheric, hydrologic, and food production cycles until they had to run all those cycles themselves in Biosphere 2.

The biospherians started to complain to each other daily about their empty stomachs. At some points, they were eating as few as 1,750 calories a day, while performing physical labor for much of the day. Each day a different crew member had cooking duty for the three meals, using the exact rations that Sally carefully measured out from the storeroom. The cooks tried to be creative with the grains and vegetables available. But after eating their small meals, the biospherians would get up from the table still hungry, Sally recalled. "It was very stressful," she said, "especially with a crew like that—essentially white middle-class, upper-middle-class Western individuals who had never been short food in their whole life—it was a tremendous shock." The hunger gnawed all the more intensely because everyone was burning so many calories working all over the Biosphere. Frustration set in as some felt themselves become mentally slower and irritable from lack of food, just when they wanted to be doing the most creative work of their lives.

The Intensive Agriculture Biome began with eighteen different plots growing different crops, but as some crops failed, the biospherians nervously stuck to the ones that proved most successful. Eventually, nearly two-thirds of the farm area was planted with the easiest three crops: sweet potatoes, lab-lab beans, and beets. The sweet potato–and-beet diet was so high in beta-carotene that the skin of the biospherians' hands turned an orange tint. (Ten years later, at least one biospherian confided that she still refused to eat beets.) Trying to use every available calorie, biospherian cooks birthed such concoctions as green papaya breakfast porridge and beet milkshakes. To supplement what they grew on their little half-acre, they got only occasional meat from the pigs and milk from the goats, and a tiny amount of fish raised in the rice paddies. The chickens laid only occasional eggs. Bananas from the agricultural area's orchard, and sometimes from the rainforest, offered one of the main sources of sweetness. They grew so abundantly that Sally wrote and performed a poem in honor of the crew's savior, entitled "Ode to Banana." The celebrated fruit's many uses included banana crepes, banana porridge, banana-goat milkshakes, stew of beans and green bananas, home-brewed banana wine. But aside from the honored banana, the wilderness areas offered little food. Coffee trees in the rainforest produced enough beans for only one precious cup per biospherian every two to three weeks. The most caffeine-thirsty crew members would pour water through the same coffee beans three or four times to eke out every drop, and even tried eating the leftover grounds.

Everyone had come into Biosphere 2 at different ages and sizes, but everyone got exactly the same rations—and everyone started shrinking. Roy dutifully weighed his companions and watched as the men lost an average 18 percent of their body weight and the women 10 percent, most of it in the first six months. The heaviest biospherian had been Taber. He entered the Biosphere at a stocky 208 pounds and dropped to a skinny 150, his old clothes hanging loosely on his bony frame. The lightest biospherian, Gaie, dropped from a thin 110 pounds to a tiny 98. When they came to the visiting windows, the biospherians looked gaunt and bony to their friends outside; the biospherians themselves, accustomed to seeing mostly each other's thin faces, remarked how bloated and puffy the outsiders looked in comparison. (However, all eight crew members regained some weight in the second year, when food intake became a bit higher.)

The biospherians split over whether this was a human experiment they wanted to be a part of; some pleaded with Mission Control to send them food. "Some of us on the inside and the scientific community were really saying look, let's take a step back, we've got problems; people are really, really hungry. They're so hungry they can't even function properly," said Jane, one of the biospherians advocating for food imports. Jane argued to her crewmates that "it makes no difference whether you produce 85 percent of your food on the inside or 80 percent of your food as long as you could calculate how much it was. . . . But the science was massively compromised because of this artificial—I think artificial—objective of absolutely nothing, absolutely no material going in and out during the two years." Her partner, Taber, agreed: "Give us a bit more food, a few more instruments," he recalled arguing. "Let's change the rules midstream because the rules we started off with are inappropriate." Taber wanted to do more chemical research, not spend his whole time fighting hunger, he said. But the image of a self-contained world had become too powerful. The directors at Mission Control refused to change their plan for self-sufficiency, and several of the biospherians themselves were determined to tough it out hungry. "Every molecule that you bring in from outside becomes part of the system, so if you import food, it would mess up the whole concept," insisted Sally, the food manager. Finally, just over a year into the mission, biospherians quietly began eating the extra stocks of dry grains and beans that they had brought into Biosphere 2 to use as seeds.

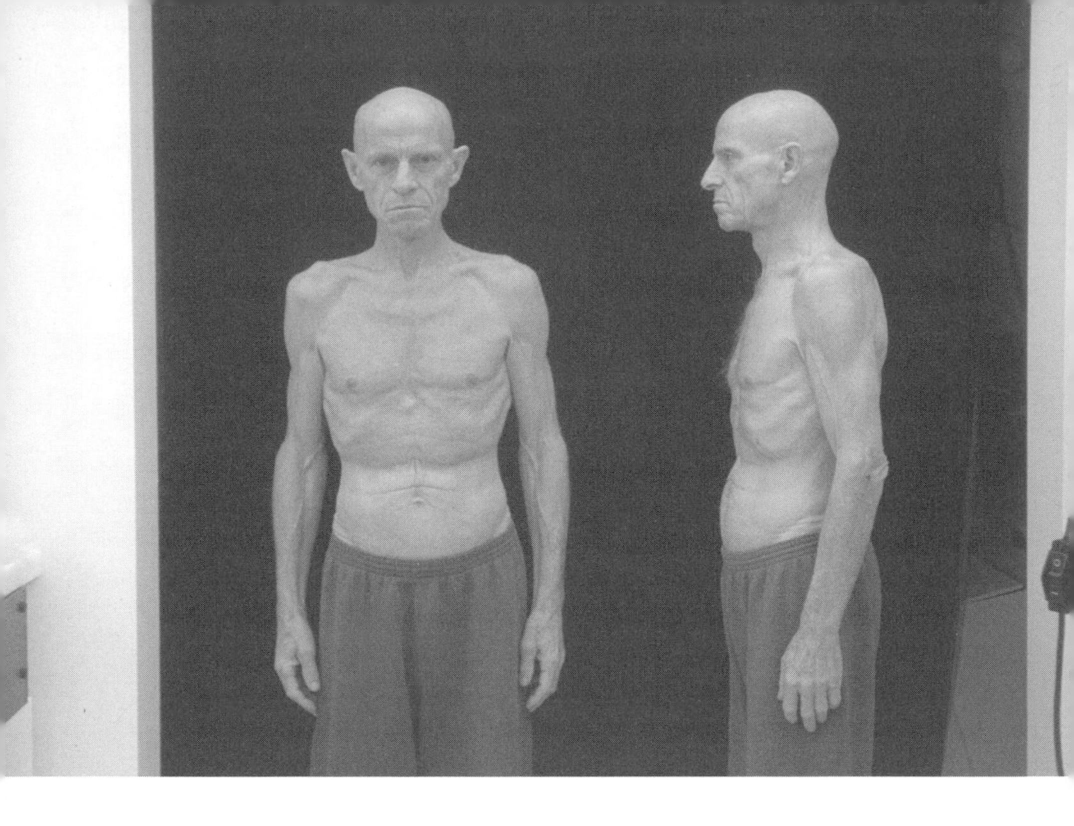

Self-portrait by Dr. Roy Walford, showing effects of the biospherians' "healthy starvation diet."

ROY L. WALFORD LIVING TRUST.

And so food became an obsession. Holding up one's dinner plate to lick the last juices was a sign of survival, not rudeness. Sally took to saving extra treats for special feasts on holidays and birthdays, when the crew would spend hours cooking together, then gorge themselves. On birthdays, the honored biospherian could order any meal. Goat cheese pizza with whole wheat crust was a favorite. Sally wrote a cookbook of her favorite biospherian recipes, to be published after the mission. At the front of the book would be a photograph of the crew at their one-year-anniversary celebration, halfway through their enclosure. They stood posing behind a table loaded with plates of food: eight identically rolled, perfect banana crepes; a huge beet salad; trays overflowing with potatoes, meat, and vegetables; pie and cake. But the biospherians standing behind the abundant food in the picture told a different story: thin smiles, sharp collarbones and chins, loose shirts.

Theater only went so far to make light of the situation. Laser, a video artist, made a short movie while inside, starring a masked figure who breaks into the Biosphere's refrigerator late at night to steal a piece of cake. In the next scene, Sally, standing in her bathrobe in the morning, screams in agony as she discovers the crime scene. But real cake was too precious to use as a prop, so the cake had to be symbolized in the movie by a block of wood. And in actuality, Sally did start to lock the food storerooms to keep would-be snackers out.

For only one biospherian, the food shortage turned out to be an oddly perfect opportunity. Dr. Roy Walford had spent years carefully restricting the diets of mice in his lab, trying to prove that limiting their food intake would extend their life spans. In the 1980s he had publicized his theories, promoting a drastically low-calorie lifestyle in books such as *Maximum Life Span* and *The 120-Year Diet.* Still, Roy had never gotten to scientifically test his theories on people because it was so hard to get human subjects to follow his strict diet in the real world. Now came the ideal research opportunity, he realized. "I think if there had been any other nutritionist or physician, they would have freaked out and said, 'We're starving,' but I knew we were actually on a program of health enhancement," the already lean Roy said. He called the biospherian food regime a "healthy starvation diet": low in calories, but high in nutrients from all the ultrafresh produce. Tracking the biospherians' blood chemistry, he found that they actually were getting healthier, according to the numbers. Everyone's blood pressure and cholesterol both dropped—just as had happened in Roy's lab mice. The numbers resembled the levels of younger people, and while Roy could not prove such a diet would actually make the biospherians live longer, he did suggest that similar diets had "promoted health, retarded aging, and extended maximum life span" in animals, when he published his Biosphere 2 findings in the high-profile *Proceedings of the National Academy of Sciences.*[6] After all, Roy's skinny lab mice outlived normal lab mice by as much as 40 percent, as his web page would later announce: "We are Dr. Walford's mice / We find a longer life quite nice."[7]

Roy also found other bizarre data when he analyzed his companions' blood: although the biospherians were living without any exposure to toxic pollutants, levels of toxic chemicals in their blood were actually rising. Levels of DDE (a by-product of the pesticide DDT) and PCBs (toxic chemicals used for various industrial purposes) both climbed in the biospherians' blood. Yet the biospherians were not being exposed to either of these chemicals at the time, and both DDT and PCBs had been illegal in the United States for years. The reason, Roy realized, was that these toxins are stored in the body's fat tissues; as the biospherians' fat melted away, they unintentionally released lifetimes' worth of stored poisons into their bloodstreams.[8]

Biosphere 2 had been created, many of its scientific advocates claimed, to learn about ecological communities. In fact, it more strikingly exhibited how humans behave in those communities when their own survival is on the line. Food shortages and atmospheric problems brought up an interesting set of land-use decisions, with striking analogies to the world outside. The ecosystems least valued by humans as "wilderness" met their demise in Biosphere 2. Tony Burgess could only watch from outside as weedy grasses encroached on his carefully created desert, as condensation dripped down from the spaceframe and biospherians poured rain on the desert to stimulate growth and take up more carbon dioxide. Fierce debates also prevailed concerning the fate of the rest of the wilderness, particularly when the crew realized how little food they would have. Sally wanted to plant squash in the rainforest, but the other biospherians vetoed the idea; the rainforest, at least, was sacrosanct as wilderness. But the status of the rest of the wild remained up in the air. John Allen instructed the biospherians to plant fruit trees in the savanna to supplement their diet. Linda, in charge of the savanna and other terrestrial wilderness areas, protested that this would be a clear violation of wilderness policy, but Gaie rebelled and led a crew of biospherians in planting the trees.

Those little battles held bigger lessons about human stewardship of ecosystems. Often conservationists and the popular media depict subsistence farmers as an enemy of the wilderness, suggesting that hungry people will always destroy wild lands to plant their crops. The biospherians proved an exception. They did alter the wilderness as they needed to meet their bare needs—for example, planting some fruit trees in the savanna—but even in the midst of stress and hunger, they poured great effort and attention into preserving the parts of their wilderness that they agreed were sacred. The rainforest in the end retained 61 percent of its original species under Linda's care.[9] The ocean too proved how strongly emotional attachment could inspire conservation. "When we closed, there wasn't a single scientist we were working with who thought the coral reef would survive," Gaie recalled. If Biosphere 2 had truly been left to evolve on its own, its water and air chemistry likely would have killed off all the corals. But through Gaie's fervent trouble-shooting, 814 coral colonies survived and 87 more were born, according to a study by collaborating scientists.[10] A shared culture and agreed-upon code of behavior usually proved stronger than hunger; the biospherians were committed to helping their wilderness survive. "Biosphere 2 became our baby, and we became as dedicated to it as a mother to a child," reflected Roy.[11] "The eight people who dwelt inside, myself included, came to regard her as 'the ninth Biospherian.' That being so, we took good care of her."[12]

The biospherians had dreamed of grand scientific discoveries. Taber had looked forward to doing pioneering analytical chemistry research, and was dismayed when labor needs in the Biosphere were so intense that he had no time for research and had to send his air and water samples out through the airlock for analysis outside. Linda too had dreamed of doing ecological research in the wilderness. "The purpose of Biosphere from my point of view was to look at how the Earth works, including human beings," Linda recalled. But, she said, in practice, "It became clear to me that I was supposed to be a laborer on the inside." Or perhaps she *was* learning about "how the Earth works," for the biospherians' lives were consumed with working out basic earthly concerns: a balanced atmosphere, clean water, and enough food. As Mark Nelson put it, "We learned more from what went 'wrong,' that is, the unanticipated, than from the multitude of things that went as planned."

Perhaps the *way* of knowing that the biospherians were developing was as significant as any actual information they received. "We all, all eight, definitely said it, time and time again, that we realized the health of our biosphere was our health. If our biosphere was healthy, we were," Gaie reflected. "And that's something that doesn't happen out here," she said, referring to Biosphere 1. "We don't live that day by day; we appreciate that knowledge, but we don't live it." For decades environmentalists have bemoaned the problem known as the "tragedy of the commons": when people do not feel the impact of their own actions, they are more willing to destroy or waste common resources like air and water.[13] It is extremely difficult to convince people to drive less, for example, because the carbon dioxide from each individual car contributes only minutely to global climate change. Individuals do not experience the negative consequences of their own driving directly and therefore feel less urgency to change their own behavior. The biospherians, in contrast, had the most intimate sense of responsibility to their atmosphere. "At breakfast meetings, we'd discuss how to keep our biosphere well so we could live well," Gaie recalled. "How to keep CO_2 down, how to try to enhance oxygen, what was going on with the oxygen, how to care for the ocean because the carbon dioxide was rising too high and the pH was falling—these basic kind of essential things that were real at the time."

The biospherians avoided using toxic chemicals not because of some abstract ideological reason, but because they knew that if they used a single drop of pesticide, or even artificially perfumed shampoo, "we would be drinking it in our tea within a week," Gaie and Mark later wrote.[14] They liked to recount the story of an open PVC glue can left behind during the Biosphere's construction. The glue fumes from an improperly closed can were enough to register on the Biosphere's air monitoring system, leading the biospherians to hunt down cans in corners of the basement. The message was clear: every action counted.

The biospherians' close relationship to their world also amplified their gratitude. Linda recalled feeling a whole new level of appreciation for her food, after laboring so hard to grow it. She delighted in counting the number of species on her dinner plate and in calculating how long it actually took to produce the rare, coveted birthday meal of goat cheese pizza—not just preparing the meal but growing the tomatoes, harvesting and processing the wheat for flour, nurturing and milking the goats to make cheese.

Interestingly, even as the biospherians paid minute attention to what they did to their Biosphere and what they received from it, a blind spot was apparent when I interviewed them: none of them seemed concerned by the massive fossil-fueled power plant that was literally sustaining their "self-sustaining" world. They paid intense attention to carbon dioxide levels in Biosphere 2, where they could feel the effects of their own actions, but the project's ecological effects on Biosphere 1 were harder to grasp, farther away. This blind spot seemed further evidence of the environmental consciousness they were already exhibiting: it was physical and emotional connection, not intellectual abstraction, that propelled people to care for their environment.

The biospherians' environmental ethic thus was grounded solidly in body, heart, and mind. They loved their Biosphere, they understood it—and in case love and science weren't enough, they knew they would have to suffer any consequences of their actions. In a speech at Biosphere 2, Carl Hodges, the agriculture biome designer, once announced, "We are all potentially biospherians." He defined a "biospherian" not only "as a person who lives in a biosphere, but rather as a person who totally understands a biosphere."[15] Would it be possible to create the same combination of cultural factors in Biosphere 1 to turn citizens into true biospherians? Biosphere 2's residents hoped so, imagining themselves as a living example to people outside in Biosphere 1. As Taber told an interviewer, the biospherians imagined that people outside would "see that those eight people can make mistakes that could destroy their world, but those eight people are being stewards of their world—and that would roll back to the people watching Biosphere 2, and they themselves would realize they are stewards of the Earth."[16] However, of course, the conceptual leap from caring about a three-acre biosphere to caring about a whole planet was a big one.

Still, while the biospherians took responsibility for their world, the most powerful creatures in Biosphere 2 were not the largest, but rather the smallest. The biospherians could tweak their atmosphere in many ways, but they were fundamentally reacting to trends the soil microbes had already set in motion. As had happened billions of years earlier in the evolution of Biosphere 1, the microbes were regulating Biosphere 2's atmospheric chemistry, which in turn influenced everything else, as the resulting changes in air and water chemistry affected plants, soils, and corals, and the biospherians themselves. The process of biospheric evolution, which the designers had eagerly looked forward to, did appear to be happening in Biosphere 2—on the microbial level. The problem was that the resulting balance was not necessarily favorable to humans and their favorite creatures. Watching the microbe-driven oscillations in the Biosphere 2 atmosphere, Tony Burgess said, the scientists began to realize, "We had not really created a model of earth, but a model of a totally different domain of life . . . It became blatantly apparent that the thing was going to evolve faster than we had planned, into Lord-knows-what." Don Spoon, a consulting microbiology researcher from Georgetown University, discovered major swings taking place in the microbial communities in the Biosphere 2 ocean when he looked at samples of its water. He even identified an amoeba previously unknown to science, proudly naming it *Euhyperamoeba biospherica*.

In the animal kingdom, the story was similar: the creatures that won out were tiny and adaptable, and functioned en masse. Out of twenty-five vertebrate species introduced into Biosphere 2's wilderness, nineteen went extinct. Hummingbirds disappeared from their little aviary in the rainforest; Linda wondered if they had been eaten by the galagos. The galagos themselves did not fare much better than their fellow primates, the humans; two baby galagos were born inside Biosphere 2, but two of the others fought with each other, sometimes violently, and one died of electrocution while exploring the Biosphere's machinery. All of the bees and butterflies, which were supposed to pollinate the rainforest, died out too. Cockroaches overran the beehives, and an uninvited bird appeared to be eating bees. For some unknown reason the butterflies never mated and died out after one generation. Meanwhile, a stowaway species of ant that had not even been invited into Biosphere 2, the tiny "crazy ant" (*Paratrechina longicornis*), thrived. Typically found in disturbed tropical and subtropical ecosystems and cities, the crazy ants loved Biosphere 2. They attacked larger insects in swarms. A later study by a group of entomologists determined that in essence, the entire Biosphere had become one enormous ant colony: out of all the invertebrates in Biosphere 2, the team found, "the only ones that now thrive are species that are 1) ant mutualists (homopterans tended by ants), 2) ant resistant (well-armored isopods and millipedes), or 3) escape ant attack by being very small and subterranean (mites, thief ants, and springtails)."[17] Cockroaches also ran rampant in the kitchen and farm, multiplying so quickly that the biospherians vacuumed them up and fed them as a snack to the chickens.

In the plant kingdom, pioneer species that were small, fast-growing, and highly adaptable also dominated. In the rainforest, tenacious morning glory vines sent out runners to climb all over other plants and the spaceframe in a tangled mat. The biospherians constantly had to cut the vines back to keep them from smothering the trees.

Eden did not seem to be cooperating gently; ecological balance could not be negotiated overnight. In the eyes of the entomologists studying the ants, Biosphere 2's plant and animal community was no pristine tropical utopia; they judged it more a model of human-altered lowlands, "a fairly good ecological analog for a small, highly-disturbed, subtropical island."

In January 1993, just over halfway through their mission, a scientist asked the eight biospherians what they wished for. They had just finished a tense videoconference meeting about their scientific plans, and now, as they sat around the conference table, they had time to relax for a moment. Some still spoke of high hopes: "To end the two years knowing what happened, to the best of our ability; that our data sets are intact and we know what happened," said Taber. Mark wished for "Big Macs and the lobster meals and the Japanese sushi and all, and the many, many alcoholic beverages that I continually dream about," but also said he looked forward to "another gigantic surprise . . . even if it makes our lives harder," arguing for the scientific value of unexpected events. (To that, Norberto Alvarez-Romo of Mission Control replied, "You don't have permission from Mission Control—no surprises!" to a chorus of nervous laughter.) The other biospherians' wishes were more humble: "That the sun comes out," said Jane, thinking of the crops. "To make it for two years," said Gaie. "I wish for eight happy healthy biospherians walking out of here in 1993 September," said Sally. "To complete all the way up to the 26th of September as a total system," echoed Laser, equally determined to get through it all. "I wish that nobody out there or in here takes their oxygen for granted," said Linda. "My wish is well known," said Roy. "I wish for a bottle of scotch."

In the second year of closure, the biospherians would regain a bit of weight; they were producing more food, and their bodies were becoming adapted to the scrappy diet. They were also seeing, said Mark Nelson, that "one consequence of the low-calorie/high-nutrient diet is that the body becomes more efficient at extracting nutrition from food." However, he acknowledged, "compared to the usual American/Western diet we were still a bit 'hungry.'"

Years later, despite all the difficulties, John Allen would claim triumphantly that the biospherians' Mission One was a success. They had proven the power of life, and that humans could channel that power, he announced in a journal article:

> The results from Mission One mean that sustained long-term inhabitation of worlds in space is possible. Mission One also means that humans can learn to collaborate intelligently with this entity that they are a part of on planet Earth. Biosphere 2 proves that life in its totality is a force tending to actively maintain and extend itself.[18]

Reflecting on her time in Biosphere 2, years later Gaie still believed that it had given her a kind of hope:

> It proved that you give life a little bit of help, and it will thrive and do well, and that's tremendous information for the crisis we live in today, that we can change what's happening, it's not too late, it probably never will be too late. It's just a matter of making a little bit of adjustments here and there and life will go forward.

Some of the Biosphere's ecodesigners drew conclusions from the project more warily, however. "Soil's alive. Don't forget that. If the soil isn't happy, nobody's happy," was one of the number one lessons Tony Burgess learned at Biosphere 2, he told an interviewer.[19] Savanna designer Peter Warshall, with a similar humility, concluded, "We are simple—a species that can hold no more than five or six variables in the mind at once. Dumping 3,000 species into a titanic terrarium is beyond our management capabilities." Reviewing the strange tasks of atmospheric management that consumed the biospherians' lives, he offered, "The lessons learned from Bios 2 have been drowned in the soap operas. Here's a simple list: Ecodesign is crucial. Surprises abound. Be humble when trying to outguess Gaia."[20]

THROUGH THE LOOKING GLASS

A DECADE AFTER THE FIRST BIOSPHERIAN EXPERIMENT, WHENEVER I TOLD anyone I was writing a book on Biosphere 2, I invariably got a common reaction: "Yeah, Biosphere 2—that thing failed, right?" In pop culture, Biosphere 2 eventually had become known as a colossal flop. A book on the "Twists and Turns of Bad Science" featured a whole chapter on Biosphere 2.[1] Comedian Pauly Shore spoofed the project in his movie *Bio-Dome,* featuring a mad scientist obsessed with "homeostasis" who tries to blow up the "Bio-Dome" with exploding coconuts, culminating in a keg party in the rainforest. In 1999, *Time* officially crowned Biosphere 2 one of the "100 Worst Ideas of the Twentieth Century."

Where did this story of failure come from? It was perplexing. Biosphere 2 did not work perfectly, but the biospherians did stay inside for the full two years. Algae were fighting for control over the ocean, grasses and shrubs were overtaking the desert, and crazy ants ran wild—but for the most part the "green slime" scenario, predicted by some critics, had been avoided. Much of the carefully designed "wilderness" indeed remained lush and alive. The biospherians had produced 80 percent of their food for two years, organically, off half an acre of land. They recycled 100 percent of their wastes, a first for closed-systems experiments. Weren't those all signs of success? The more I investigated the "failure" narrative, the more instructive I found it, but not because of any inherent truth it contained. As a whole, mass media accounts of Biosphere 2 revealed little about Biosphere 2 itself, but they told plenty about the relationship between society, science, and nature in Biosphere 1.

At first, the media bought into the excitement of Biosphere 2–as-new-frontier. Magazines and newspapers displayed photos of the smiling crew in their matching red jumpsuits. Journalists, like the biospherians themselves, got hooked on the notion that the biospherians, by stepping through a door, could escape to a faraway planet. A 1989 report in the *Arizona Daily Star*, covering Gaie's five-day stay in the Test Module, described "a revolutionary Test Module that is, for all practical purposes, *completely isolated* from the outside world"—even though the article ran alongside a photograph of a hundred guests and reporters crowded around the little glass house to peek in at its human resident, hardly a picture of "isolation."[2] When it came time to seal Biosphere 2, the *Arizona Republic* headlines declared the "Biosphere crew's 'last day' on Earth."[3]

Public fascination with Biosphere 2 followed in a long American tradition of survival spectacles. Stories of extreme explorers have long appealed to people mired in comfortable civilized lives. In 1913, in a primitive parallel to the Biosphere 2 story, the American media became obsessed with the story of Joe Knowles, a man who abandoned all his clothes and tools and wandered into the Maine woods to see if he could survive for sixty days in the wilderness.[4] A late twentieth-century version of the survival-spectacle genre, contemporary with Biosphere 2, was the story of a young man named Chris McCandless, made famous by the book *Into the Wild*. In the early 1990s, McCandless, a young college graduate from a seemingly normal family, determined to begin living "*Real* life," gave away his possessions and savings, and hitchhiked to Alaska to live in the bush, where he later died.[5] The story struck a nerve in the public, becoming a best-seller. Moreover, in researching McCandless's story, author Jon Krakauer came across several other examples of men who had done nearly the exact same thing: young white men from middle- to upper-class backgrounds had repeatedly abandoned their lives and chosen to put themselves on intense survival missions out in the wild.

Thus the modern attraction to wilderness adventure arose mainly among people surrounded by civilization. As historian William Cronon observed, "Ever since the nineteenth century, celebrating wilderness has been an activity mainly for well-to-do city folks . . . The dream of an unworked natural landscape is very much the fantasy of people who have never themselves had to work the land to make a living."[6] The ultimate American wilderness survival icon, Henry David Thoreau, exemplified that contradiction. "In wildness is the preservation of the world," he famously wrote. His account of life in his cabin at Walden Pond became the bible on American self-reliance. But in fact Thoreau's "wildness" adventure would have been impossible, physically and metaphysically, without the nearby town of Concord. The town provided him with supplies and visitors, and even more crucially, the town's civilized existence was what inspired him to try life in the "wildness" in the first place. By the end of the twentieth century, the American public lived more distant from nature than ever. Yet people continued to display a voyeuristic attraction to wilderness survival stories, as though yearning to understand a dimension of life unavailable among the comforts of society. These popular stories ranged from best-selling books about Antarctic expeditions to "reality TV" shows such as *Survivor.* Biosphere 2 was part of that pattern of spectacles. Americans were curious to see whether the biospherians could survive in a wilderness of their own creation. A typical magazine account of Biosphere 2 summarized what the public believed were the project's main goals: "In 1991, eight men and women locked themselves inside an enormous glass-and-steel shell in the mountains north of Tucson to see if they could survive for two years with no outside help."[7]

In fact, many of the Biosphere's founders did not originally want media attention—or so they claimed. "We had hoped to keep it a secluded enterprise and were surprised at both pro and con treatments" by the media, John Allen told me. For years the creators had traveled among their own projects, able to operate according to their own ideas of synergetic civilization. "Within this context they were living and working with unconventional people, and could pretty much make their own rules," recalled Kathy Dyhr. "I am not criticizing that," she insisted, as she had loved that life too; "but when the Biosphere project became high-profile and in a sense mainstream, suddenly there was pressure for John and Margret to act according to mainstream rules. Well, they never wanted that in the first place." But public relations consultant Terrell Lamb, who later joined the project, pointed out the obvious: "When you lock people up in a bubble, you're going to get some press attention."

Rumblings that Space Biospheres Ventures was not exactly an

orthodox corporation began in the local Arizona press around 1990. That story went national in April 1991, when closure of the Biosphere was just six months away. "Take This Terrarium and Shove It," declared the cover of the *Village Voice*, New York City's liberal newsweekly; "The Media Loves It, Yale Loves It, Phil Donahue Thinks It's Neat, The Smithsonian Lends Its Name, Scientists Take Its Money—So What if the Biosphere 2 Is Really Run by a Wacko Cult? Don't You Want to Go to Mars?"[8] In a long-winded exposé, writer Marc Cooper condemned Biosphere 2 as a pseudoscience sham run by a cult. He blasted other journalists and scientists for believing in Biosphere 2, dug up stories about Synergia Ranch from the 1970s, and called John Allen "much more the Jim Jones than the Johnny Appleseed of the ecology movement." The Biosphere-builders tried to fight back. John Allen typed up a scathing seven-page letter to the *Voice*, calling the article "an utterly fallacious and willfully malicious attack against myself, endangering my economic livelihood." It was "false that I ever 'ran a Gurdjieff group' or any other kind of cult group," he insisted. But the damage had already been done. Smelling a scandal, national media picked up the *Voice* story, and by the time the biospherians walked through the airlock, the project's formerly glowing reputation had been tainted.

With the biospherians inside, press coverage quickly spun out of control. "I got a phone call saying, 'These reporters are all calling us, they're asking us if we're part of a cult,' and suddenly I was in the midst of this," recalled Terrell Lamb, a PR professional who was brought in from the East Coast to help stem the damage. She remembered the night she arrived in Arizona; intense people were bombarding her with questions from every side. "This press campaign . . . it was really twenty-four hours a day," she said. "And all kinds of press—the *Washington Post*, the *Fort Worth Star-Telegram*." Terrell had a long career in public relations, and she compared the media circus at Biosphere 2 to working on a high-pressure New Hampshire presidential primary.

Rumors popped up all over the headlines. Anyone who had ever felt slighted by John Allen started turning up in the newspapers. Anthropologist Laurence Veysey's reports on Synergia Ranch, based on observations made twenty years earlier, found their way into numerous newspapers and magazines. Reporters dredged up old allegations, first printed in the mid-eighties in a Fort Worth paper, that John Allen had physically and verbally abused his companions, including Ed Bass. "It was always 'the cult' and the media pressure from the outside saying how weird everything is," said Safari, who worked outside Biosphere 2 during the biospherian mission, consulting with the biospherians on animal and agriculture issues. Safari herself had been cut from the biospherian team at the last minute, because Margret Augustine worried that news reports were mentioning that too few biospherians had scientific degrees, and Safari didn't have one. "Media were lurking left and right; you didn't know if you pick up the phone if it's somebody from the media disguised," Safari recalled. "We were getting pretty paranoid about everything. We didn't know what to say and how to say it, it would just be blown up right away."

Biosphere 2 had started out as a media darling, heralded as a "new kind of world." What had gone so disastrously wrong? Part of it was simply poor public relations. Journalists were retaliating with exposés because they "knew they were being lied to," said biospherian Taber. For years, putting on the theatrical performance of a big science project, no one working at Biosphere 2 had dared mention the staff's after-hours lifestyle to the press. "There was absolute fear that the outside community would find out that we had that kind of structure," said Linda. When the secrets started to come out, the press assumed it all must be scandalous, she said. "When people guessed there was this structure out there, they thought it was something really bad and evil." Fearing that media spies were everywhere, the synergetic team installed air-conditioning units in the dinner room so that they could have communal dinners with all the doors and windows closed, recalled Jane Poynter. Meanwhile, reporters got suspicious at simply the aura of abnormality. Terrell Lamb, the PR consultant, remembered watching science writers befuddled by John Allen's metaphysical ramblings and speeches. "If you're a science reporter, you're very used to 'This is the process, this is where we're going'—it's very black and white," she said. "Johnny, when he gets going, you have to have an appreciation for who he is to really get it, and most people don't, because he gets in your face, he gets really intense." Journalists "found the people that they met to be 'odd,'" she said, and "stopped giving them the benefit of the doubt."

The biospherians had hoped to catch the public eye, but in a very different way. By putting their lives on stage, they believed, they would offer the world an inspiring model for ecological living. In a society full of people cut off from nature, Biosphere 2 would "give them images of harmony between people and living systems," said Mark. Linda wrote in a column for an environmental magazine, "Because of the high profile [of Biosphere 2], we can get the point across about the importance of that potato you eat. Or the importance of: Where does the water go? Or what happens when you cut your front lawn?"[9] At least, that was the hope.

However, the rest of the world turned out to have different ideas about what "self-sustaining" meant. To the biospherians it might mean living in ecological harmony with their world, but the press became more concerned with the theatrical aspect of the experiment as a survival stunt. "Planet in a bottle," read a *Washington Post* magazine cover devoted to Biosphere 2 in January 1990—but the media appeared much more interested in the idea of a sealed "bottle" than in the comparison to the planet. Tens of thousands of tourists flocked to Biosphere 2. In the first six months of the biospherians' mission, 159,000 people—nearly one thousand every day—took guided tours around the Biosphere's perimeter, peering through the glass, hoping to glimpse the elusive residents of the human zoo going about their daily lives. As one newspaper reporter observed after taking the public tour, most tourists' curiosity was not exactly the kind of attention the biospherians had been hoping for: "Visitors seem most interested in finding out why anybody would volunteer to be sealed away for 24 months, and what *exactly* are the sleeping arrangements."[10]

The press did take interest in the idea of Biosphere 2 as a new world, but harped on one aspect: it must be a sealed, self-contained world, where nothing could come in and nothing could come out. And so began the allegations that the biospherians were "cheating." The first incident concerned the installation of a carbon dioxide "scrubber" machine in Biosphere 2's basement just before closure. In the months before final closure, when the managers sealed Biosphere 2 for brief test runs, CO_2 levels had shot upward. Just two and a half weeks before closure, Taber, who would not only analyze the Biosphere 2 atmosphere but breathe it, finally got permission from his bosses to install carbon dioxide–scrubbing equipment. The new machine could remove excess carbon from the atmosphere as air passed through a sodium hydroxide solution, which removed CO_2 by a chemical reaction. The machine's output was a dry powder that could be stored in Biosphere 2. Later the CO_2 could be released back into the Biosphere's atmosphere by heating the powder in a furnace—though in actuality CO_2 levels would stay so high that this last step would never be performed, while the basement filled with barrels of the scrubber's by-product.

When a few news reporters found out about the scrubber machine's unannounced last-minute installation, they cried of a scandal. National media quickly seized on the story. "Some say this belies claims that Biosphere 2 would rely solely on natural, biological activity to cleanse both air and water," reported the *Boston Globe*.[11] The *Tucson Weekly* was more blunt: "The Biospherians Cheated When They Installed an Extra Air Treatment System," it announced in a headline, and argued that the scrubber installation was "in gross violation of the underlying principles of the experiment."[12]

The notion that a new technological addition to Biosphere 2 somehow violated the "principles of the experiment" was, of course, ridiculous in light of how much technology was already keeping Biosphere 2 alive—from air conditioners to rain machines to wave pumps to the huge generators powering it all. Indeed, for some of the designers, the whole point of Biosphere 2 was to integrate ecology and technology for nature's benefit. Architect Phil Hawes, frustrated with the critics, believed that the discipline of "biospherics" must include "not only the biology and its ecology but also the technology required to keep the system alive," he said. But journalists seemed caught up in the illusion of Biosphere 2 as a purely wild Eden, even as the rainforest hooked up to pipes, wires, and fans stood as a cyborg right in front of them.

The media caricature of Biosphere 2 as a hermetically sealed bottle became even sharper a month later. Just two weeks after the biospherians locked the doors behind them came the first accident. While Jane was threshing rice using a machine in the basement, she sliced off the tip of her finger. Medical officer Roy quickly went to work sewing the fingertip back on. In a video he took of the surgery, Jane lay on the operating table singing operatically and shouting as he stitched her finger back together. But when Roy told her she might have to leave the Biosphere for follow-up surgery, she bolted upright on the operating table, tears welling up in her eyes: "You mean I'd have to go out? Couldn't we just fix it up so it can last for two years?! That would really be disastrous if I had to go out . . . I couldn't do it." The biospherians' dedication was so intense that the patient seemed more upset about the prospect of leaving Biosphere 2 than she did about her own severed body part. Soon after, she was standing at the airlock door, forcing an actress's cheery smile for the cameras and stepping into an ambulance. Five hours later, Jane was raced back from a Tucson clinic, her hand wrapped in thick bandages. However, as she stepped back through the airlock, her other hand held a mysterious duffel bag full of unnamed supplies. Media critics went crazy with speculations of what might be in that bag. "For want of a fingertip, might an experiment be lost?" demanded the *Arizona Republic*.[13] The *Boston Globe* revealed the press's skewed idea of the project's ecological goals, arguing that the CO_2 scrubber and the finger incident "deviated from the experiment's original claim to be a totally closed and self-sufficient system."[14]

The Biosphere-builders were partly to blame for setting themselves up for such treatment. They had created high expectations in the first place. "No air, water or material will cross the airtight boundary of space-frame and glass between Biosphere 2 and the surrounding biosphere of Earth," announced one of their typical press releases. But journalists seized on that aim as though it were the only goal of the project. "Imports may tarnish project's successes," announced the *Arizona Daily Star* in a final evaluation just before the biospherians were to emerge. More than half of a newspaper page meticulously listed every item that had been imported through the airlock over the course of two years—mainly small pieces of electrical equipment, film, medications, office supplies, and tiny predatory mites for agricultural pest control—as though this was the most interesting information to come out of Biosphere 2.[15] As ecologist Howard Odum noted, "some journalists crucified the management in the public press, treating the project as if it was an Olympic contest to see how much could be done without opening the doors."[16]

Living in a glass house, the biospherians literally had nowhere to hide from the media onslaught. At one point a news helicopter even buzzed around their apartments, sneaking photos. The biospherians felt misunderstood and attacked. "It was insane that we had that much animosity from the media, because we were working our guts out inside on a really very worthy, worthwhile venture," Jane said. Biospherians spoke to reporters on the phone and at the Habitat visiting windows, and on special occasions answered questions by videoconference. But the stories that made it into print were overwhelmingly negative. "We'd have these interviews with the press . . . and they'd put out the next day everything other than what we said," Gaie recalled. "The media was just going haywire on us, and not only saying we were leaving the Biosphere because we couldn't breathe, gasping for breath, to saying we were cultists to God knows what, and here we were, believing entirely in what we were doing." She believed with her whole being that "this was an historic mission," she said, but she could only look out the window and wonder: "Can anyone say we're just doing OK, can someone give us a pat on the back, please? Is there anyone out there, except for the immediate people working on this project that are all under attack, that can extend a hand of friendship, and just appreciate the glory of it all?"

As reporters looked for any flaw that they could turn into a news story, they were creating a Biosphere 2 narrative only marginally resembling the plot lines happening inside the Biosphere itself. The word *media* is a telling term; newspapers and other news outlets literally mediate between their readers and the real world, and in the process create their own dramas. Those created dramas become almost real when they are repeated over and over in news outlets throughout society. "Journalism is storytelling," as media scholar Gabriel Weimann describes it. "The integration of news and entertainment, facts and fiction, events and stories" creates "a symbolic environment in which fact and fiction are almost inseparable."[17] The biospherians had chosen their own archetypal dramas to give their quest meaning—space, the frontier, the new world, Eden, ecosustainability—while the press eventually settled on a simpler story line about cheating and failure, and made that the socially prevalent version of the Biosphere 2 story.

To the heads of Biosphere 2, those media attacks felt like volleys in a war. Kathy Dyhr, heading public affairs, watched her bosses, John and Margret, become more and more secretive and guarded in reaction, she said. "It was a siege mentality. They were up against the wall." The media accounts of Biosphere 2 became so powerful—especially when repeated among so many news sources—that they became a self-fulfilling prophecy. The press had decided that "if Biosphere 2 isn't perfect, it's a failure," lamented biospherian Mark Nelson—but the Biosphere's managers seemed prisoners of the same logic. They stopped referring reporters to speak with savanna designer Peter Warshall because, Warshall said, "I would just tell them what had gone on . . . but the management always felt it reflected badly on them if I said, 'Oh yeah, seventeen of thirty-five introduced species are now dead.'" Even people working at Biosphere 2 were kept from knowing the full extent of what was going on inside; if they did not know about problems, they could not leak them. Taber, the analytic chemist inside the Biosphere, was the first to discover the Biosphere's steadily declining oxygen levels. John and Margret insisted he keep quiet about it at first. When John later reported the oxygen problem to the project's group of scientific advisors, he included a graph predicting that the oxygen drop would stop of its own accord, as though everything were under control. Tom Lovejoy, the biologist chairing the project's Scientific Advisory Committee, described the mentality of his contacts at Mission Control: "The problem was in the end they got hooked on a vision of it all working perfectly, whereas from a scientific point of view, that was totally unnecessary."

As part of its story of Biosphere 2–as-failure, the media homed in on another theme as well: was Biosphere 2 doing real "science"? Journalists appeared to believe that there must be a clear-cut answer to that question. In response, the Biosphere's managers struggled to conform to public expectations of what science should be. The ensuing debate, again, would not teach the world much about what was really happening at Biosphere 2, but it would illustrate the near-mythic status of science and scientists in contemporary society.

In reality, many leading practitioners of ecology and engineering were advisors to the Biosphere 2 project, informally or formally. John Allen could proudly rattle off a list of his most esteemed friends and consultants who visited or advised the project: microbiologist Clair Folsome; brothers Eugene and Howard Odum, founders of systems ecology; biochemist Harold Morowitz; Oleg Gazenko, head of the Institute of Biomedical Problems in Moscow, and other top Russian scientists studying and creating artificial habitats for space; British geophysicist Keith Runcorn, a Fellow of the Royal Society; and others. Some of these men sat officially on the project's Scientific Advisory Committee.

Yet all this expertise did not seem to matter much to the press when the charge that Biosphere 2 was "unscientific" began, with caricatures and name-calling. Marc Cooper, the author of the *Village Voice* tirade that originally turned the media tide against Biosphere 2, devoted pages to slamming scientists for working on such a "decidedly nonscientific" project. Cooper appeared to have in mind a specific definition of what was science and what was not. "The biospherians may be *talking* science, but what they are *doing* is more akin to well-financed science fiction," he wrote. Moreover, he implied that the separation of "science" from "science fiction" was about morality, as though science were a religion and those violating its tenets were heretics. For example, he argued that the involvement of high-profile researchers from the Smithsonian and the New York Botanical Garden in designing Biosphere 2 "*does* raise serious and disturbing questions about the ethical standards of the U.S. scientific community."[18]

Biosphere 2 thus became a battleground over the unwritten rules of what did and did not count as "science." Name-calling critics launched their assaults: "unscientific," "pseudo-science." The *Tucson Citizen*, which had glowed over "a New World" in 1987, now warned in late 1991, "While crew members are calling it a scientific experiment, scientists across the nation have called it just pseudo-science."[19] One reporter actually asked Biosphere 2 PR representative Chris Helms, "Is Biosphere 2 science or science fiction?"[20] Writers—and, perhaps, their readers—seemed to assume not only that "science" could neatly be separated from "nonscience," but that the project's worth would be determined by which side of the boundary Biosphere 2 fell on.

Those assumptions had surprisingly deep sources. Mentions of "pseudoscience" were actually historically loaded accusations. Morally weighted arguments over the meaning of "science" go back to the foundations of modern European science, when Protestant Reformation leaders denounced alchemy as heresy, as though science was a new religion, as cultural critic Andrew Ross has pointed out. Looking at the history of Western science, Ross argued, the demarcation between science and pseudoscience "is *always* historically specific, determined by the cultural and ideological circumstances of its day, and thus by the particular claims that 'science' and 'scientists' make for themselves in a particular time and place."[21] Therefore, to understand contemporary attitudes toward nature, it is more interesting to leave aside the question of "Was Biosphere 2 science?" and instead ask, "What does such a question even mean about the society that asked it?"

Several popular assumptions about science were on the line in the war over Biosphere 2's credibility—and indeed, they did reflect cultural and ideological circumstances of their time. "Science" could be performed only by official scientists; only the right high priests could interpret nature for everyone else. "Science" was separate from art (and, by implication, the thinking mind was separate from the emotional heart). "Science" required some neat intellectual boundary between humans and nature; it did not necessarily involve humans learning to live with the world around them. Finally, "science" must follow a specific method: think up a hypothesis, test it, and get some numbers to prove you were right. Those assumptions all exploded like bombs in the biospherians' faces, for they had violated the unwritten rules of how scientists interact with the natural world.

It is worth unraveling each of these assumptions about science, for these beliefs, which surfaced repeatedly in critical attacks on Biosphere 2, reveal much about the relationship between science and society in Biosphere 1, particularly in American culture. Beginning with the first assumption, in journalists' eyes, "science" depended centrally on what kind of people were doing it—were they real "scientists"? Many journalists tried to answer this question by reporting basic facts: there were few advanced academic degrees among the eight biospherians. Linda had done some graduate science coursework, Gaie had a master's degree in environmental studies, and Roy was an MD, but no other biospherians had official scientific credentials, in spite of their many years of Ecotechnics work. Some had not even been to university. Defending themselves, the biospherians pointed out that they spent hours on the phone with their professional scientific consultants outside. Distinguished ecologist Howard Odum also defended them:

> The original management of Biosphere 2 was regarded by many scientists as untrained for lack of formal degrees, even though they had engaged in a preparatory study program for a decade, interacting with the international community of scientists including the Russians involved with closed systems. The

history of science has many examples where people of atypical background open science in new directions.[22]

But criticisms of the biospherians as nonscientists went further than just their lack of diplomas. Here came into play the journalists' second assumption as well: art and science may not mix. As news articles dredged up stories of the Theater of All Possibilities, journalists held up the theater troupe's very existence as proof that the biospherians were not scientists, as though it were impossible to be an actor and a scientist at the same time. A television newscast used stories of Synergia Ranch to discredit Biosphere 2: "The commune members took part in strange costumed rituals . . . Hardly anyone at the ranch was a trained scientist. They were actors in the Theater of All Possibilities." The show pointed out that "Johnny Dolphin," theater director, was now director of research and development, and that "Margret Augustine, or 'Firefly,' went from wardrobe mistress to chief executive officer and coarchitect of the Biosphere."[23]

The Dolphin and Firefly scurried to hide their theatrical past. One interviewer asked Margret Augustine—who remained not only the project's CEO, but its theater director—if it were true that "on weekends, the group closes up shop and gets together for dramatic training." She cut him off: "Space Biospheres Ventures is a business . . . Our business offices are closed on Friday at 5:30, and then in terms of after that, we're a business, that's the weekend, and we start up again on Monday."[24] The denials only made journalists hungrier for the truth. A *Life* magazine writer reported what happened when she asked whether closed group psychology would be part of the Biosphere 2 project: "'We have only so many resources,' I am told, and my attention is directed back to 'science.' 'Look at the science,' they keep saying." But, the reporter noted, "The more the biospherians say 'science,' the more I am reminded of the Wizard of Oz: 'Pay no attention to the man behind the curtain.'"[25]

The great irony of this scramble to claim Biosphere 2 as strict science was that such a position ran entirely contrary to the builders' most heartfelt philosophical aims. The founders of Biosphere 2 had always believed that their work was about much more than science. They knew that any successful project must have strong group dynamics at its core. Since the 1970s, the Biosphere's board members—John Allen, Marie

Harding, Margret Augustine, Mark Nelson, Ed Bass, Kathelin Gray—had been working together to develop themselves into artists, scientists, and explorers. Kathelin recalled that she and her fellow actors had always wanted to be "spiritual and secular at the same time," to become like traditional societies—such as in Bali, Indonesia—where theater, science, and the sacred were one, all "integrated with everyday life." But now that idea suddenly seemed blasphemous to media critics.

At the beginning, the Biosphere project had embraced and straddled art and science. At one of the earliest planning conferences in 1985, theater director Kathelin, known as "Honey," sat alongside NASA astronauts Rusty Schweickart and Joe Allen for a panel discussion on the human side of Biosphere 2. Honey introduced herself as "a cosmonaut of inner space," and all three shared stories of their travels and experiences working in groups. At another early scientific planning meeting, the theater performers put on a skit for the conference, dancing around on the lawn, pretending to have turned into wild animals living inside the Biosphere; Ed Bass romped around like a pony. As the visiting scientists watched, "Everybody was completely in culture shock at this point," recalled Tony Burgess, laughing. But the scientists were willing to accept their offbeat new bosses, he said; after all, it was clear that "this is a delightful, wacky group that's got a lot of money behind it."

The builders themselves accepted that Biosphere 2 was, in a way, their biggest theater production. "I heard Johnny saying that the whole Biosphere is actually just theater; it's a theater performance," said Bernd Zabel. The biospherian candidates all participated in theater, and Kathelin helped them write their speeches for the Biosphere entry ceremony. As Kathelin put it, she and her colleagues believed Biosphere 2 would include art and science together; in "systems science, unity, totality, trance-like connection with the environment," there would be no separation. But when the media declared, "It's not real science," she felt like she had to hide, as though "the theater, the artistic side wasn't even there," she said. "It affected all of us. I felt like I couldn't really say who I was."

Why must art and science be separate? Why must the thinking mind be separate from the feeling heart? The answer to that question lay in the historical roots of modern science. During the Scientific Revolution of the sixteenth and seventeenth centuries, scientific men enshrined an experimental, matter-based approach to knowledge, in fields ranging from physics to philosophy. Sir Isaac Newton described the world in terms of mechanical laws. René Descartes declared, "I think, therefore I am," privileging rational thought (as opposed to emotional feeling or sensory experience) as the foundation of human existence. A cultural change was under way. Earlier Europeans had imagined themselves as part of nature, an organic whole in which they lived; but during the Scientific Revolution, through the lens of science and technology, they began to view nature as an object, separate from people, to be manipulated and controlled. Historian Carolyn Merchant described this change in detail in her book *The Death of Nature*. As Merchant argued, that change would have lasting results: "The mechanistic view of nature, developed by the seventeenth-century natural philosophers and based on a Western mathematical tradition going back to Plato, is still dominant in science today."[26]

Another historian, Peter Dear, has called the distinction that arose during the Scientific Revolution a divide between learning by "experience" and learning by "experiment."[27] One who learns by *experience* becomes integrated into his or her surroundings; one who learns by *experiment*, in contrast, seeks to separate him or herself from the "experiment," manipulating and observing it from a distance. The biospherians were learning by personal experience. Meanwhile, they lived in a time when science was supposed to require a detached, objective observer. The biospherians' very existence as members of their own experiment confounded the categories. As Margret Augustine described in one presentation, "the inhabitant is necessarily a manager and researcher; further . . . the manager is necessarily related to the wealth he or she produces, and the researcher is part of the experiment and studies which he or she conducts."[28] The biospherians understood their world by discovering, day by day, how to live with it. As Gaie put it, most of what they learned "was practical and managerial—in fact all of it was, because we had to keep the thing running."

Here entered the third critical assumption about science that was wielded against Biosphere 2: apparently, learning to live in harmony with the surrounding Biosphere was not "science," according to the critics, including many in the scientific establishment. A *Life* article on Biosphere 2, entitled "Weird Science," printed a typical criticism from a Yale botany professor: "So much of Biosphere 2 is just bass-ackwards. In science you ask a little question to get the answer to a big question. They're not asking a question; they're saying, 'Suppose we build this thing? What's going to happen?' That's a question, in a way, but it's not science."[29]

Because Biosphere 2 did not have a "little question," was it therefore not science? Here entered the fourth critical assumption about science: science must follow a precise set of steps, generating a narrow hypothesis, testing it, analyzing the data, and producing a definite answer. Prominent ecologist Howard Odum, a staunch advocate of the project, maintained that Biosphere 2's emphasis on "using data to develop theory, test it with simulation, and apply corrective actions was in the best scientific tradition." He argued that critics unfairly assumed Biosphere 2 to be unscientific simply because its complexity could not be immediately understood by scientists used to operating on smaller scales: "When journalists asked establishment scientists, most of whom were small scale (chemists, biologists, population ecologists), they got back the dogma that system-scale experiments are not science."[30]

In fact, the debate over Biosphere 2 had become a debate over the meaning of science itself. Biosphere 2, as a megaexperiment, was being judged by the criteria of other scientific disciplines. The "Big Science" of hundred-million-dollar scientific facilities, in America, had been devoted to pursuing narrow questions in physics, in order to meet U.S. military demands; now Biosphere 2 was judged as though it must be in the same category of experiments.[31] Fields like physics and molecular biology focus on highly controlled, single-variable experiments to test simple hypotheses. Schoolchildren memorize from an early age that the "scientific method" consists of a neat set of steps: hypothesis, experiment, data collection, data analysis, conclusion. But Biosphere 2 had never been designed for that kind of science in the first place. As desert designer Tony Burgess argued, "We had made the decision early on that given the nature of a complex process, we were quite comfortable, as the biologists involved, as good people, we were going to watch the system structure itself, collect as much data as we could, and see if we could figure out a pattern after the fact . . . It's good natural history." In Biosphere 2, the simple hypothesis-testing mode of inquiry would have made little sense, Burgess argued: "It seemed very premature to ask a detailed question or propose a detailed hypothesis for testing, when you're just trying to figure out how to build a system." Indeed, as biospherians Mark and Gaie freely admitted, when they stepped into Biosphere 2, "not only did we not have all the answers—we didn't even have all of the questions yet!"[32]

Many Biosphere 2 science consultants had embraced that uncertainty, including well-respected scientists who had made their careers in more controlled styles of experimentation. Ecoengineer Carl Hodges eloquently proclaimed to a gathering of his fellow Biosphere designers that they must break free of science's normal reliance on monitoring a few "key variables":

> If you grant me the premise that led to our being gathered here together—that is, that a true biosphere is a whole interactive "living entity"—then *every* variable is a *key* variable, and to list "key" variables to the exclusion of "non-key" variables would be to regress to a paradigm that, almost by definition, would not allow us to succeed. That is, with the state of knowledge today, we cannot build Biosphere 2 from fundamental equations upward.[33]

Likewise, Roy, the biospherian doctor, had spent his life doing orthodox controlled experiments on lab mice, but he too saw value in approaching Biosphere 2 as "a voyage," he argued in the scientific journal *BioScience*. In the journal, Roy declared the travels of Charles Darwin to be a model for "serendipitous science." Darwin set sail as a naturalist on the *Beagle* without knowing that he would develop insights into the theory of evolution: "He didn't know exactly where he was going, nor what he might expect to find there, and he had *no hypothesis to test*. He was merely a thinker-observer—of course an astonishingly good one—on a long, eventful voyage."[34] Similarly, Biosphere 2 might yield unpredicted insights, if its observers were able to receive them, Roy believed.

But would a naturalist such as Darwin even receive funding in a modern scientific environment? As Howard Odum would later argue, two schools of ecological research existed by the late twentieth century. One approach, aimed at generating numbers and provable, simple facts, "isolates parts of the environmental system for intensive study, experimental testing where possible, with hypotheses relating a few variables." The other approach "combines information about many parts and relationships in order to infer how the various parts work, how the many connections and relationships operate together, and how the system might be changed to solve problems that have been identified."[35] C. S. Holling dubbed these "two cultures of ecology" as the "analytical" and "integrative" approaches to science.[36] But one of those two approaches controls the bulk of research funding—and it is the approach focused on control and simple facts, not the approach focused on complex interconnections and human participation in ecosystems. An entire academic economy has grown up around scientists pursuing short-term, hypothesis-testing research projects, funded by short-term research grants. Scientific journals publish mainly short papers based on some single analytic insight. Research grant cycles, typically lasting a few years, encourage scientists to stick to quick hypothesis testing—as does the academic tenure cycle, which depends largely on the number of papers a researcher publishes. Biosphere 2 had originally been exempt from those limitations because of its massive private funding. The biospherians knew well, said Jane Poynter, that "if there had not been private funding, it never would have happened . . . It's not the standard way of doing science, and you're not going to get grants to do that kind of thing." However, the Biosphere's private funding did not make it exempt from being judged publicly by people used to a different kind of science.

As a concession to the critics, Biosphere 2's managers tried to adjust to speaking the accepted language of science. Over time, the glossy *Biosphere 2 Newsletter* turned toward more articles on scientific aspects of the project. Several of the Biosphere's creators argued in the biology journal *BioScience* that Biosphere 2 "is intended to help evaluate the hypothesis that a biosphere regulates its environment by means of an ecological feedback process involving its microbial, plant, fungal, and algal communities."[37] Elsewhere they argued that Biosphere 2's purpose was "to test the hypothesis of adaptive self-organization."[38]

Beyond that broad "hypothesis," however, the Biosphere's builders themselves did not know exactly what they were looking for. "Shall we write a history of atomic migrations, of genetic change, of ecosystem configurations, of agronomic production, or of human awareness?" Tony Burgess asked flat out, during a presentation in 1987. The project's open-endedness resulted in a massive data collection effort. The biospherians amassed piles of information, sending it out for analysis and advice. A system of two thousand computerized sensors kept track of temperature, air and water quality, and soil moisture. Taber regularly sampled Biosphere 2's air and water and sent them out through the airlock for chemical analysis. Scientists at institutions across the country pored over records of the Biosphere's cycles of carbon dioxide, oxygen, and trace gases. Meanwhile Linda kept detailed notes on development of the Biosphere's wilderness plants and consulted with the biome designers outside. In the marine environment, Gaie collected samples for a host of researchers who were helping to study topics ranging from ocean chemistry and coral vitality, to small-population genetics of *Gambusia* fish in the marsh, to microorganism biodiversity in the seawater. Even within these studies, the research setup "wasn't traditional," Gaie said. "Instead of having to write a proposal to NSF, each one of the scientists could come to us, and if we were able to take on a project, we did it."

Still, amid those early mountains of data, it was not yet clear what Biosphere 2's great scientific achievements would be. John Allen had hoped to discover the "laws of biospherics." However, much of the knowledge about Biosphere 2 remained locked inside the biospherians' minds. In spite of all the computerized sensors accumulating data, Gaie said, "The biggest learning of the Biosphere was the eight people inside . . . actually where the data is held is in us, more than anything—and Johnny on the outside who also was just living it; he tracked that twenty-four hours a day. And that's probably also why the operating manual of the Biosphere also was never built, because it was in us; we never had time to write it out."

While the biospherians learned from experience, they remained unsure how to translate that data into something the outside world could learn from as well. "The human couldn't take all that data and learn from it," Gaie said. "You could make presentations of data, but it never achieved the ability where you could take the nerve system of Biosphere 2 and jack in, 'Aha, this is how healthy the Biosphere is.' You could only glean bits of information from all that data." The biospherians may not have been doing conventional science, but if they were on the way to inventing a new way of doing science, that science was not yet fully defined.

All the criticisms of Biosphere 2 as "cheating" and "unscientific" at last came to a head in a bizarre twist. Oddly enough, one of the biggest supposed "failures," in terms of the project's goals of maintaining a closed system, turned into one of its biggest successes as "science"—illustrating the degree to which notions of both "failure" and "science" were heavily socially constructed. The problem began as a puzzle. Oxygen was going missing in Biosphere 2. While the carbon dioxide level rose steadily, it actually was not rising as much as it should have, mathematically speaking. Engineers could calculate how much O_2 had gone missing from the atmosphere, but it was not all showing up in the form of CO_2. Some unknown source in Biosphere 2 appeared to be sucking up carbon dioxide.

No one on the Biosphere 2 staff, and none of their scientific consultants, could figure out what was happening. They began to look for outside opinions. One night in New York City, John Allen met with well-known Columbia University geochemist Wallace S. Broecker. Broecker recalled how Allen warily flashed him a few graphs, fearful of telling anyone about Biosphere 2's woes. Broecker, who studied geochemical cycles in Biosphere 1, was intrigued, and agreed to come out to Arizona for a look. In the end it was actually one of Broecker's graduate students, Jeffrey Severinghaus, who solved the mystery. Jeff's dad, a building engineer, suggested to his son to check out the concrete inside Biosphere 2—as it cures, concrete absorbs carbon dioxide. The biospherians drilled out cores from their concrete walls, cliffs, and mountains and sent them out for chemical analysis. As it turned out, the concrete had not finished curing before Biosphere 2's closure, and the CO_2 in the air was reacting with the Biosphere's internal walls and fake rocks to form calcium carbonate. By sucking up CO_2, the concrete was actually saving the biospherians and other creatures from even higher potential CO_2 levels.

The mystery had been solved, but that did nothing for the biospherians' breathing. The oxygen level kept plummeting. The crew noticed themselves panting as they walked up stairs during their daily routines. A year into their mission, Roy, Jane, Linda, and Taber began waking up at night gasping for breath, and started sleeping under masks hooked up to tanks of oxygen extracted from the Biosphere's air. As the oldest member of the crew, Roy felt the low oxygen's effects most strongly.

His heartbeat sped up, his voice became slurred, and he wobbled as he moved. As medical officer, he was in charge of telling Mission Control whether the biospherians' health was in imminent danger, but he wondered how he himself would be able to respond if a medical emergency arose. Finally he realized he was becoming mentally impaired from the lack of oxygen to his brain. "I was in the medical facility with Taber, for some reason I was adding up a column of numbers, not some complicated job . . . and I couldn't add them up," Roy recalled. "I was a physics major at Caltech; I could add up a column of numbers." Oxygen levels had fallen to 14.2 percent of the atmosphere, equivalent to the oxygen content in the air atop a 17,000-foot mountain.

A few months into the biospherians' second year inside, it was impossible to go on with the goal of a totally closed system. In January 1993, trucks injected liquid oxygen into Biosphere 2, enough to raise the atmosphere to 19 percent oxygen. Appropriately, the infusion of fresh air was made at the Biosphere's domed West Lung. The biospherians stood in the tunnel outside the lung door, waiting. When they entered the rich air—26 percent oxygen in the lung itself—they felt instantly healthy, euphoric. "It was a sense of intense well-being to breathe deeply and not feel like I needed another deep breath immediately thereafter," wrote Linda. "Barely noticing the people outside looking in at us through the lung window, I got a totally unexpected and sudden impulse to run around the lung for no conscious reason, just an impulse which drove my legs." The other elated biospherians started jogging after her, in circles around the big round room. "I felt like a born-again breather praising the virtues of oxygen to my formerly breathless companions," Linda wrote.[39] Just minutes earlier, they had been panting from climbing up the stairs.

The most unexpected result of the oxygen problem was neither scientific nor health-related: strangely, the oxygen problem helped to repair Biosphere 2's battered reputation. Jeff Severinghaus published his analysis of the oxygen mystery in *EOS: Transactions of the American Geophysical Union*.[40] In news articles, previous critics of Biosphere 2 now announced that the oxygen loss "shows they're conducting a real experiment," as one NASA scientist put it.[41] "Project's woes may aid science," the normally critical *Tucson Citizen* announced.[42] Years later, the Biosphere's builders themselves strove to frame the oxygen problem as a success. John Allen argued that the oxygen decline was not a problem but an interesting "experiment of 'riding the oxygen curve down,'" and that "discoveries about the effects of oxygen and the ability to operate at less than 20.9% may be used in the design of bases in space, submarine, and mountain work."[43] William Dempster, the Biosphere's chief engineer, argued that the oxygen decline was a great event because it proved that Biosphere 2's engineering was a success: "Biosphere 2 was so tightly sealed that it made it possible to see that the oxygen loss was occurring," he said, and this "was one of the milestones that Biosphere 2 represents." Indeed, Biosphere 2 leaked only 10 percent of all its atoms per year. The number might sound high, but in fact it was extremely low—the total equivalent of only a two-centimeter-wide hole distributed over the surface area of the entire structure. In comparison, NASA's closed ecosystems leaked 10 percent per *day*. "In Biosphere 2, once we realized oxygen was missing, we could look for it," Mark Nelson reflected proudly. Thus, years later, he called Biosphere 2's biggest problem "a great victory." To conform to traditional conceptions of "science," it seemed, you did not have to achieve ecological balance and harmony; what was most important was that you could prove you had learned something. As Mark put it,

> Learning what is out of balance is as important as learning what is in balance since it deepens our understanding of how to modify system designs in future. What was also interesting about the oxygen decline was that it was not predicted by anyone . . . underlining just how little we understand about why biospheres work and how much is to be learned before such systems can be made with high confidence that they will operate reliably.[44]

In other words: searching for the cosmic, the biospherians found the most mundane of answers, quite literally, in the concrete.

THE HUMAN EXPERIMENT

"I'VE GOT THOSE NO CHOCOLATE FONDUE, NO HOT WATER BUBBLE BATH, wallowing algae scrubbers, back-breaking, total process breakdown blues," crooned the four women of Biosphere 2—five months before they actually entered the Biosphere. In a play they wrote, directed, and performed for a crowd of their friends and coworkers, titled *The Wrong Stuff* (a spoof on *The Right Stuff*, the book about NASA astronauts), the eight biospherians had acted out, in exaggerated hysterics, all the crises that might befall them in their new world. Wearing red and blue wigs, painted faces, and glittering costumes, they pranced about the stage colliding with each other, pulling on then–crew captain Bernd Zabel's limbs until they threatened to tear him apart, arguing over whose area of the Biosphere deserved the most attention. They rolled on the floor to a drumbeat, moaning and chanting in unison, "I'm sick, I'm sick, I'm sick of you all." In one scene, Sally meted out food to each biospherian for a meal: three lettuce leaves to one, two radishes to another, a kumquat to a third; sobbing and panting in hunger, the crew sang, "One more Swiss chard, and I am going to puke; that one good meal, well it was just a fluke." In another scene, the crew members panicked because they were supposed to write a scientific report. "All we need is a lot of numbers, a lot of graphs, and just make it look beautiful. We'll call it 'creative data'!" Jane exclaimed, and they all began furiously scribbling nonsensical graphs on a screen. Later the eight split into two groups of four, shouting at each other in a "battle of the sexes."

Though John Allen was never mentioned by name, his presence was inferred in a scene toward the end. "Look at this team. How could we ever succeed with a team of eight people like this? Sitting in his office, writing another scene for his new play," Linda snarled. Sally chimed in: "Nobody can operate this catastrophe because you designed it that way, and now you're out there and we're in here!" In anger, the crew mutinied against "Mission Out-of-Control," chanting, "Cut the cable, cut the crap!" The play ended with the biospherians strutting about in a song and dance number that concluded in their forming a human pyramid—which collapsed to the ground. The audience of their friends and coworkers laughed hysterically and cheered.

And yet every single one of those scenarios that the biospherians acted out, in some variation or another, took place in Biosphere 2.

Before he locked himself inside the Test Module for the first time in 1988, John Allen had asked for words of wisdom from the Russian scientist Evgenii Shepelev, the first man ever to inhabit a closed ecological system decades earlier. Shepelev responded, "Remember, man is the most unstable element in the system." This was the irony of Biosphere 2. In their efforts to manage every aspect of an ecological system, from ocean chemistry to atmosphere to agricultural production, the biospherians discovered that the least manageable factor of all was the one seemingly closest at hand: the human mind.

This was the final dimension of the human experiment: while the biospherians worked hard to take care of their Biosphere, their own human relations fell apart. Within six months, the crew had split into two factions. Sally, Mark, Gaie, and Laser—agriculture director/crew captain, director of communications, marine scientist, technology wizard—were on one side, following the instructions that John Allen phoned in from Mission Control. Roy, Linda, Jane, and Taber—doctor, wilderness ecologist, animal caretaker, chemist—were on the other side, sick of the Biosphere's outside management regime and sick of their crewmates' loyalty to it. Each foursome included one male-female couple—Gaie and Laser on one side, Jane and Taber on the other. And each side believed deeply, even years later, that they were right. Living without enough food and enough air naturally made people more irritable. But the divide went deeper than disagreements and bickering; it turned into a tribal loyalty. "Four of us were not on the side of the management, and four were devout followers of John Allen," said Roy. Jane, Taber, and Linda had all worked on Biosphere 2 since its genesis and participated enthusiastically in the synergetic life (Roy had joined later to become the doctor); but now those four felt alienated from their one-time leader and irreconcilably split from their other four crewmates. "At six months we all kind of tried to pretend it wasn't there," said Jane, regarding the split. "And about the year point it became clear that we couldn't pretend any longer."

Disagreements flared into snide comments and insults. "It was as if we were starting the process of divorce after a passionate love affair had gone awry, each party feeling betrayed by the other," Jane later wrote.[1] Social interactions deteriorated to the point that at some times the two halves of the crew barely spoke with each other, except when absolutely necessary for work. At feasts and birthdays they still managed to be civil to each other. But often they walked by each other coldly in the high-ceilinged Human Habitat, as Jane wrote: "One afternoon, Taber and I were walking along the hallway to the dining room. Gaie and Laser were walking toward us. As we passed them we hugged the wall, and averted our eyes. So did they. That was the way it was for the remaining fourteen months. We never looked each other in the face again."[2]

Nevertheless, the eight kept working together. They had devoted their lives to Biosphere 2 and were determined to make it succeed. "I don't like some of them, but we were a hell of a team," said Roy. "That was the nature of the factionalism . . . but despite that, we ran the damn thing, and we cooperated totally." Mark was in the opposite faction, but he agreed: "The more dangerous part of an isolated group is the psychodrama," he said, but in Biosphere 2, "It was such an evident reality that Biosphere 2 was our life support. Nobody sabotaged the lifeboat . . . There was this tender loving care and symbiosis." Yet all the loving care and symbiosis directed toward plants and nonhuman creatures did not spill over into relationships between the eight humans. The situation quickly became demoralizing. Only two months into the two-year mission, Linda wrote in her journal, "This is the only situation I have ever been in that would drive me to drink, except there is no drink in here."

There was nothing to do except keep working. The doors were closed. The biospherians had worked so hard to get to that point. Now they walked by the airlock every day, but no one dared walk out the door. "We were completely and totally stubborn about the fact that we were going to be in there two years," said Jane. "I certainly was not going to be the first one to say, 'I'm out of here, kids, bye.'" Meanwhile, the biospherians quietly hid their rift from the public. "When I was preparing for the mission, I was wondering if I would have to deal with an endless number of petty squabbles," Sally told a local news reporter, "but people really made quite an effort not to push each other's buttons."[3] "All through the two years, when we were not talking to each other inside for eighteen months, we were stating to the outside world, 'Hey, we're getting along like a house on fire, we're just having a great time,'" Jane recalled. "I never quite understood the rationale for that." At one point, Tony Burgess was touring a group of students around Biosphere 2. He stopped at the Human Habitat window, where biospherians could communicate by telephone with visitors on the other side of the glass, to introduce them to his friend Linda Leigh. "Linda, tell these students what it is that they would really need to be trained in, in order to work inside the Biosphere," Tony said. She replied, "They must learn to work with people they despise in order to get a job done."

In their split, the biospherians were not unique. They were manifesting a common facet of closed group psychology. Studies of confined isolated groups, such as NASA astronauts and Antarctic research teams, have revealed a perplexing tendency: such groups nearly always split. After coming out of Biosphere 2, Taber and Jane, both of them space enthusiasts, visited several NASA centers for debriefing by government psychiatrists. NASA researchers wanted to understand the biospherian experience because they had witnessed so many interpersonal problems among astronauts once small groups were stuck together on a ship far out in space. At each NASA center, "they all said, 'You guys are textbook,'" Taber recalled. "There's no record of an isolated group, like Antarctica, of six or more, that didn't split into subgroups or cliques within." Furthermore, many studies of isolated group psychology report a mysterious "third quarter phenomenon": conflicts become the sharpest during the third quarter of the enclosure, no matter what the length of confinement. The biospherians reported a similar experience; they bickered the most just over halfway through their mission. Jane put it bluntly: "I actually got a shrink halfway through, because it was getting too weird. I actually thought I was losing my mind at one point." To which Taber quipped, "Several of us did." As one way of coping with their alienation from their crewmates and bosses, Jane and Taber began using fax and phone lines to build their own house in Tucson, planning their life together for after Biosphere 2. The title of a chapter in Jane's book summarized her experience: "Starving, Suffocating, and Going Quite Mad."

Why is it so hard for people to get along, even when they share such common goals? There is not an easy explanation. One NASA engineer has suggested that part of the problem in many small group expeditions is that the crew members are subject to many potential sources of stress: adaptation to a new physical environment; new restricted diet; a high-pressure work situation; constant interpersonal interaction; and lack of privacy. These physical and emotional stressors combine to create a psychological pressure cooker, in which "strained crew relations, heightened friction, and social conflict are expected," the NASA engineer concluded.[4] Most of those same outer-space stress factors were equally present at Biosphere 2. Certainly the intense physical circumstances did not make it easier to get along, as Jane wrote:

> Were we all mad, or was it just me? . . . as I looked at my other fellow inmates, I was sure that some of them were also quite ill. A couple of biospherians showed up for the early morning chores less and less often. They seemed listless and ate meals by themselves. Others seemed rigid with fury. Several had grown exceedingly thin and drawn. It was getting scary.[5]

For years, Institute of Ecotechnics members had hopped around the globe from project to project, creating and running from one adventure to the next. When they were entrenched at Biosphere 2, however, there was nowhere to run when things got tense. "When you're in there for a short period of time, you can put off everything—it's like 'OK, I'm just going to muscle through,'" said Jane. "But when you're in there for a certain period of time where suddenly you define it as long-duration, you have to deal with your own psychological problems . . . you have to start dealing with your own personal baggage, whatever it is." Laser, from the opposite faction, agreed: "It gave me insights just into living with other humans, which is not at all different from living with other humans outside a closed system. It was just that it was so much more concentrated that you couldn't get away from the observations." Maintaining harmony in a small closed group proved as challenging as maintaining balance in an artificial ecosystem.

The biospherians' conflicts also mirrored other isolated group missions in another way: their disagreements revolved around the role of Mission Control. In studies of both Antarctic and astronaut teams, many of the biggest interpersonal conflicts have pitted the crew against their bosses outside—a common enough scenario to be informally dubbed the "Us versus Them Syndrome." As one researcher on confined group psychology noted, "the crew . . . assume a responsibility for themselves and their environment. As this happens, they begin to see the 'outside authority' as unnecessary and conflict can occur if the central authority pushes itself on the crew."[6] In fact, in a 1984 speech at the first Biosphere 2 planning conference, Apollo astronaut Rusty Schweickart had tried to warn the Biosphere-builders of that very problem. He cautioned them that many people already had experienced the group dynamics of isolated expeditions—in Antarctica, in space, in the deep seas—and it was "not always happy." He told the story of the *Ben Franklin*, a sea lab that floated below a ship in the Atlantic Ocean. The crew was living in difficult conditions deep in the sea, but "the ultimate decision-making authority resided in the ship." When the ship authorities and the deep sea crew got into power struggles, the ship directors even threatened to cut off the sea lab's food supply, Schweickart recounted. But, he told the Biosphere 2 team, "I'm sure you won't make that kind of mistake."

The biospherians had heard such stories, had studied group psychology, had acted out the scenarios of the group's splitting or mutinying in their own original play—yet still they reenacted those stories. Like the deep sea lab floating far below the ship in the Atlantic, the troubles inside Biosphere 2 were closely linked to troubled relations with the managers outside. The biospherians were physically isolated by thick glass, but they were hardly independent. The white arches of the Mission Control offices sat only a stone's throw away from the sloping glass walls of the ocean and rainforest. In the Theater of All Possibilities, John Allen had always been the dramaturge, writing the play from behind the scenes, and the same became true at Biosphere 2.

He was determined to steer the evolution of this complex little world. That did not sit well with four of the biospherians. At one point John wanted more rain on the rainforest, which he believed would stimulate plant growth to take up more carbon dioxide. Linda, the botanist living inside with the plants, believed he was using an engineer's equation-oriented brain and did not know enough about the health of the plants to know how they would actually respond. So she defied Allen's rain-making instructions and then lost her power. On orders from Mission Control, Laser, captain of the "technosphere," locked the rain control boxes so Linda could not set the timers.

It was not coincidental that the administration building outside Biosphere 2 was called Mission Control. At times, interventions by John and his coleader Margret went beyond atmospheric chemistry. The two "would get very angry at you over any disagreement . . . anything that came down from the outside management was supposed to be followed, even if it was nuts, and sometimes it was," said Roy. A decade after Biosphere 2, when I met Roy, he was the only biospherian still on speaking terms with all seven of his former crewmates; he was the elder statesman of Biosphere 2, the doctor and avant-garde artist, who had plenty of practice in intense situations. But too often, he said, he felt sucked into conflict with his fellow biospherians and Mission Control. One day the eight biospherians were sitting around their granite dining table having a meeting, and Roy wanted to audiotape the session for his records. Sally, the crew captain, objected. Roy suggested they put the issue to a vote; it was seven against one, with Sally in the minority. So Sally called her good friend CEO Margret Augustine to complain. Margret promptly ordered Roy that he could not tape. "The people in Mission Control outside were intervening in interpersonal conflicts inside the Biosphere," said Taber. "They were causing more friction because if you said something to somebody, you would hear it from somebody else, via Mission Control outside." Linda observed that though Biosphere 2 was designed in the name of ecological "self-organization," its social structure was never allowed to self-organize.

As I interviewed the eight biospherians—four of them still living together at Synergia Ranch, the other four living different lives in Arizona and California—I was struck by what drastically different mental universes the two foursomes inhabited. They had shared their physical world as intimately as eight people could, not just working and living together, but eating exactly the same food, breathing the same air, facing all the same situations every day for two years. Yet, listening to them describe their relationships, it sounded as though they might as well have been living in different biospheres. Trying to explain to me what happened in Biosphere 2, each biospherian seemed to imagine himself or herself as the tragic hero in a theater production gone wrong. Their narratives did not sound like facades; each one believed deeply in his or her particular version of events. But there were two very different versions.

Four biospherians acted perplexed, even offended, that anyone would have defected from the side of John Allen. "Any time you're in an organization, and there's a boss, whether you like it or not, there's a boss, and in the case of research and development, Johnny was boss," said Gaie. As for the other biospherians, she said, "They can gripe about it, bitch about it, moan about it, as much as they want, and maybe they agree with nothing he did, but he was the boss, and he was the boss from day one." Nearly a decade after her exit from Biosphere 2, Gaie's eyes still lit up when she talked about the latest inspiring idea she had discussed with Johnny. Intelligent and enthusiastic, she had become one of the leaders of Ecotechnics projects. Sally, when it came her turn to sit on the Synergia Ranch couch and talk with me, was also upset about the defections. The antimanagement crew members were like traitors, she said:

> Some of us were absolutely aghast that certain factions of the crew could just turn around and turn on these people, who frankly had got them everywhere that they were at that point, and who'd given them every advantage, every opportunity, every support, to do what they wanted to do in life, to be part of a fabulously creative project, being given this privilege that was given to eight individuals on the entire planet, and then they turn around and literally bite off the hand that fed them.

The situation was "devastating," Sally said. Mark Nelson agreed: "Imagine NASA astronauts deciding they were tired of Mission Control instructions and the overall plan for their space mission!"

Meanwhile, from the other camp, Jane wrote of John Allen, "For ten years, I had believed him infallible, held him in such high regards that nobody could have lived up to the godlike image. When . . . he showed himself to be chipped, cracked, less than perfect, humanly flawed, my adoration turned to hate."[7]

On ecological issues, the two sets of biospherians gave me roughly matching explanations of what had gone right and wrong inside Biosphere 2. But when we discussed the human social issues, each half of the crew had a distinctly different idea not just about the cause of the problem, but even about what events had literally happened. The way the two cliques described their crew's decision-making structures exemplified this difference. Some biospherians reported a Biosphere run by orderly due process, aided by a wealth of support not only from Mission Control but from helpful scientific advisors at far-flung academic institutions outside. That was Sally, the crew captain's, account: "Decisions were made by a very democratic process, actually. When problems came up, we would report what we saw, and then there was a whole team of people on the outside in different areas who would take that problem, and discuss it and look at it, and make suggestions and make decisions about what strategy to try first and second and third." She went on to describe how as farm manager, she received support and advice from a variety of expert agricultural consultants, such as Richard Harwood at Michigan State University, who continued to do research to improve Biosphere 2's crops. Gaie told a similar story: "We had a lot of help. There's a lot of information out saying how all of us weren't PhD scientists and we weren't doing any science. In fact, we had some of the best scientists in the world working with us." Gaie explained, for example, how she took underwater videos of corals in Biosphere 2 and sent them to marine biologist Phil Dustan at the University of Charleston, South Carolina, for analysis.

The four alienated biospherians, however, told a different story. Jane had helped develop and design Biosphere 2's agricultural systems, but once inside, "I was not even allowed to call my agricultural consultants," she said, in contrast to Sally's account of openness and support. Said Linda, "Four of us, including Roy and myself, who had most of the scientific basis of the Biosphere, were stripped of decision-making, and at one point I wasn't even allowed to talk to certain scientists in my field."

Media coverage of Biosphere 2 had long since spun out of control, but the project leaders still were desperate to control information leaking out from anyone they did not trust (including certain biospherians). A year into the biospherians' mission, Mission Control issued official lists of which Biosphere 2 staff could talk to outside scientific consultants in which fields. According to her original assigned duties, Linda was supposed to be in charge of Biosphere 2's terrestrial wilderness areas, whose design she had coordinated, but soon she found that she was on neither the "systems ecology" nor "biogeochemistry" communication lists, which consisted of Gaie, the engineer Bill Dempster, and Norberto Alvarez-Romo at Mission Control. Her professional phone calls from scientists outside were funneled to Mark Nelson.

Whether or not Biosphere 2 would "succeed" or "fail" in its first try, its leaders seemed determined to complete a harmonious dramatic performance. In early 1992, they hired Chris Helms, a seasoned PR professional in Tucson, to repair the Biosphere's battered reputation. Helms at first thought the offer was a joke—Biosphere 2 had the worst press he'd ever seen—but he considered it a new challenge. Several months later, in November 1992, the biospherians were so hungry that they began to eat their emergency stocks of agricultural seeds as food. They agreed to tell no one. But, in desperation, Jane made a phone call to Chris Helms. Jane told him "We're starving . . . we have to have some help," Helms recalled. Feeling upset, Chris went to Mission Control to confront Margret Augustine and John Allen. Immediately they yelled at him: "Who told you?!" Helms refused to reveal his informant, but eventually Jane confessed. Margret Augustine called Chris Helms back into her office. She became enraged, screaming, and told him he could no longer make outgoing phone calls without going through Mission Control first—even though Chris's job was public affairs. When he threatened to resign, she finally backed down, he recalled; his skill in building relationships with the local press was desperately needed.

As for the traitorous Jane, Margret "fired" her from the crew and suggested that she leave Biosphere 2 under some invented story such as pregnancy. But when a good enough excuse could not be concocted, she stayed. "People were fired and unfired so often that it was almost as if John and Margret considered firing merely an extreme form of ordering someone to go stand in the corner," Jane recalled.[8] As she was doing her chores in the animal bay and then walking up the stairs one morning, loyalists Gaie and Laser, one at a time, retaliated by each spitting in her face. "Laser's spittle was especially offensive as he was chewing on peanuts. Jane ran crying to her room," Roy recalled. "Gaie later apologized; Laser never did."

Soon after, Margret Augustine also told Linda that she could leave Biosphere 2 if she was unhappy. At one meeting, Gaie suggested that she would rather send one or two (unnamed) biospherians out rather than import food. But still they were all so committed that no one wanted to leave their beloved Biosphere, even as it pushed them all to their physical and emotional edges.

The biospherians had plenty of practice working on group dynamics, but now their old modes fizzled. At first the crew kept up their old traditions of putting on skits and practicing theatrical exercises, but eventually these dissolved as some biospherians stopped coming to theater sessions. The crew went back and forth over whether they would follow the decades-old synergetic traditions of Thursday night discussions and Sunday night speeches. Some biospherians took to collecting their food and going off to eat in their own rooms. One of the biospherian factions—Gaie, Laser, Mark, and Sally—later wrote a journal article describing their attempts at maintaining group culture: "We set up special weekly events, such as Sunday night dinners to exchange personal anecdotes, stories, memories, or revelations as a way to celebrate the completion of the week. This simple social form/task has great efficacy bringing the team together because each person shares a personal part of themselves." However, they blamed the other four biospherians for keeping the program from working: "Unfortunately, some members of the Biosphere 2 team ceased coming to this occasion late in the closure experiment, which may have resulted in reduced social bonding."[9] Whatever the cause of that "reduced social bonding," the old social forms, which had held the synergias together for so many years, were no longer succeeding.

Yet art still provided a refuge. Roy organized the "first interbiospheric arts festival" in which the biospherians used a video phone hookup to exchange performances with poets and musicians at a café in Los Angeles. Roy and Laser both shot film and video, capturing both serious and silly moments. Taber and Jane had brought a collection of musical instruments inside, and enjoyed jam sessions across the wires with their friends who lived in apartments on the Biosphere 2 site. One of their masterpieces was the music video *Ecological Thing*. In it, they and Roy and Linda, dressed in matching red shirts and red helmets, marched in lockstep through the dry savanna, climbed along the glass walls, and danced, grinning maniacally, on the library's glass floor. "Ecological thing, what are you going to bring to girls and boys?" rang out Jane's voice above synthesized beats. The video showed them hurling a huge pair of dice in the ocean and on the rainforest floor, while their voices intoned over and over, "We're gonna roll the dice! We're gonna roll the dice!"

One of the wildest was Roy, regardless of the fact that he was nearing age seventy. One day he got Taber and Jane to join him in "going native"; they took off their clothes, covered themselves with body paint, and pranced around the rainforest's little pond, hidden among the plants. Another time Roy and Taber, the oldest and youngest crew members, got into a mud-wrestling match in the rice paddy in the agriculture area, while a crowd of tourists looked on through the glass.

During time off one day, Roy united the crew to paint a giant human mural on a blank white wall inside the Human Habitat. The biospherians gleefully stripped off their clothes and covered themselves with paint. They lifted each other up and pressed each other's skinny naked bodies against the wall, printing a huge chariot shaped out of painted human body parts. Clusters of blue-painted butt-prints formed the hubs of the wheels; blue legs were the spokes radiating outward. Lengths of red arm-prints, joined at the hands, formed a carriage above. Jane's and Taber's crouching silhouettes, spray-painted green, were the beasts of burden pulling the vehicle. The crew entitled their creation "The Butt-Wheeled Wagon of Biosphere 2." "If NASA pooped out, this was to be our wagon to the stars," Roy later wrote. The art project was created in a rare unifying moment of paint-covered fun, he recalled. But Margret Augustine heard about the mural and ordered it be taken down two weeks before the end of the biospherian mission. Gaie and Laser scrubbed their teammates' bony butt-prints off the wall.

Birthday parties and seasonal holidays such as solstice became rare happy times when the crew put aside their squabbles and came together to feast on the meal of the birthday biospherian's dreams. Copious amounts of food helped to ease the tensions. To create the illusion that they were going out, the crew held their party picnics all over the Biosphere—sitting on a blanket on the tiny ocean beach, on couches in the library tower looking out at the mountains, on the balcony overlooking the farm fields below, or even at the conference table in the computer-filled command room. Everyone had brought in colorful costumes, and they partied as drag queens and space aliens. One night they got drunk on home-brewed banana wine and danced their hearts out on the command room tables, while Roy videotaped the spectacle. Other times they just lay back on pillows on the floor after a rare satisfying meal, bloated and content. But outside of those moments, during the routines of daily life and work, the fundamental rift kept coming back.

Meanwhile, the discord inside Biosphere 2 was mirrored in the staff outside. "The day-to-day operation, it just became awful," said Safari, who worked outside the Biosphere on domestic animal and agricultural development. An aura of fear surrounded John Allen and Margret Augustine, who appeared to be under great stress. Margret became "stormy, tempestuous," computer technician Gary Hudman recalled; "everyone was frightened to death of having to talk to her."[10] Even approaching the cream-colored Mission Control building felt dangerous to some. The building's steps "were like gold," said Kristina Sanchez, an office employee; "you were honored to walk up those steps." Gilbert LaRoque, a tour guide on site, recalled being told, "If you go up these steps, you'll automatically be fired." At times, when Margret wanted to walk from her office to her house at the other side of the Biosphere campus, which would require her passing between the research greenhouses and the gift shops, Gilbert would get calls on the radio asking that he clear all tourists out of the area because Margret didn't want to see anyone.

The all-consuming quick pace from the construction days continued. Work was "every day, nonstop, seven days a week, morning to night," said Safari, and "I didn't have any problem with that, but I didn't like the social stress of the infighting, absolutely not, and I just thought, I don't want to live that way." She finally left the project while her friends the biospherians were still inside, because "I just couldn't handle all the fighting any more," she said. "Margret was the CEO, John Allen was sort of really in charge, but then there was a lot of fighting between them, disagreements between them . . . it just started totally crumbling apart."

Why a once-solid social structure had so completely imploded would continue to consume the Biosphere-builders' thoughts for years afterward. Various people speculated different reasons to me: Was a whole miniworld just too big for anyone to manage? Was there a power struggle inside the management? Was the real problem the added stress from constantly being criticized by the outside world? "Generally the commune . . . kind of would keep to themselves and function alright," said Roy, "but when the Biosphere was being built there was all these scientists around, and the press was around, so it was not going to keep by itself any more. So that put it under a great deal of strain, and the system started falling apart." Somewhere along the line, creative energy had turned destructive. There was no easy explanation. Linda voiced a vision that came to her mind: "A couple of us had a feeling of John as the Anti-Christ, who was coming back to destroy what he had created."

Somehow the dynamic of "create and run" did not translate well into "create and stay." John Allen himself seemed to acknowledge such a dynamic years later in his memoir: "All my life, since I first left home at fourteen . . . my motto had been 'create and run.' Continuous creation was my dream. My attitude, until well after the Vajra hotel had been finished, had been as if my life had been symbolized by Brahma and Shiva alternating—creation and destruction. I never really went for the Vishnu super program, the maintenance force, until later . . . just before Biosphere 2 started."[11]

Amid the tumult, the Biosphere's benefactor, Ed Bass, remained quiet. Usually Ed played a hands-off role at Biosphere 2, spending much of his time back home in Texas. But he apparently was worried about the project's scientific reputation. Ed had asked eight high-profile and well-credentialed scientists to join the project's Scientific Advisory Committee (SAC), including the Smithsonian's Tom Lovejoy, a famed conservation biologist, as chairman; Gerald Soffen of NASA; systems ecologist Eugene Odum; and Biosphere 2 rainforest designer Ghillean Prance, who by then was director of the Royal Botanical Gardens at Kew, England. The committee's aims were scientific but also theatrical: to improve the Biosphere's science program and to rebuild its reputation. In early 1992 Ed announced that the committee would report to him on ways to improve Biosphere 2's science. In the process, the scientists would become drawn into a power struggle.

The committee of prestigious scientists visited Biosphere 2, and talked to researchers and biospherians. After some months, in July 1992, the SAC issued its long-awaited report. While acknowledging that Biosphere 2 had much scientific potential, the committee recommended that it needed a stronger, more specific science plan "which sets research priorities for ongoing and future projects, includes detailed project budgets, states long- and short-term goals, and outlines methods, anticipated results, and potential significance." However, John Allen did not interpret those suggestions as helpful recommendations. "The review committee insisted that there be a hypothesis and a research plan. He just went absolutely ballistic. He said, 'This is an attempt to control,'" Kathy Dyhr said. Tony Burgess recalled the same: "John at this point said, 'Ed's trying to get rid of us; he's going to use this Scientific Advisory Committee to get rid of us, so that the establishment can take over this project.'"

Perhaps that impression was not far from the truth. Though its mission was supposedly to investigate Biosphere 2's science, the advisory committee turned around and started on an investigation of the project's management instead. Some committee members began prodding Ed Bass to get rid of John and Margret. Committee secretary Rob Peters began making phone calls to Biosphere 2 employees, asking them for criticism about Margret Augustine, John Allen, and the third-highest officer at Mission Control, Margret's husband, Norberto Alvarez-Romo. One long-time staff member wrote a secret memo to the science committee, accusing John and Margret of "explosive outbursts of rage; inappropriate argumentation and tirades against staff, professional associates, even occasionally journalists; erratic and arbitrary decision making; inability to tolerate a contrary opinion much less criticism," which were becoming "an across the board response to the external world."

Meanwhile, John Allen worked to defend his terrain. At one Biosphere 2 board meeting, he suggested that the members of the Scientific Advisory Committee, as they were not employees of the project, not be told any information that could be considered "proprietary"—and then went on to suggest that *all* scientific information about Biosphere 2 might be considered proprietary.

At last, just when the biospherians were breathing fresh air after the infusion of oxygen in January 1993, the management threatened to combust. It began with a biospherian calling out for help. Tony Burgess got a phone call from an upset Jane, repeating what she had told Chris Helms: the biospherians were very hungry and were eating their agricultural seed stocks, but they were not supposed to tell anyone. Tony felt worried and confused. Like everyone else, he had to choose sides. When his turn came to testify before the Scientific Advisory Committee, Tony recalled, John Allen gave him a pep talk like a coach before a football game, encouraging him to go in and convince the committee that Biosphere 2 was on the right track. "It just turned my stomach," Tony said. "I thought, wait a minute, this whole thing is theater." The next day, in a closed-door meeting, Tony sat down with the committee and unloaded his worries about the biospherians' physical and mental health. It was not just hunger, he told the scientists; he worried that some were psychologically near breakdown, having to turn to counselors outside Biosphere 2. The committee members fired questions at Tony, trying to get the dirt on the management. "I was surprised at how personal some of the questions were," Tony said. "I answered the ones I thought were appropriate." Then the panel of scientists thanked him, and he got in his car and drove away, believing that he might be leaving Biosphere 2 for the last time. When he saw John Allen a few weeks later, John greeted him with a growl: "Betrayal is punished by the lowest depths of hell."

The session caused an explosion. Argumentative phone calls and faxes flew back and forth between Biosphere 2, Ed Bass's office in Texas, and the committee leaders. Finally the committee fell apart. Ed Bass fired its secretary, Rob Peters, and other members hastily resigned. "It became very hard for us to fill our role" was the only comment that chairman and conservation biologist Tom Lovejoy would offer me about his resignation. "If we had felt that we could make recommendations about the science that would be appropriately entertained, we would have stuck it out," he said. The public dissolution of the committee—despite the fact that five of its nine members continued to consult individually as scientists at Biosphere 2—shot another wound into Biosphere 2's scientific reputation. "Will anything come of it? So far this is just a stunt," Gerald Soffen, one of the committee members, bitterly told the press.[12]

Biosphere 2 was supposed to be an icon for ecological sustainability. But another question loomed as time went on: what about economic sustainability? Construction costs had mushroomed from original $30 million estimates to more than $150 million once Biosphere 2 was standing. Operating costs totaled an extra $50 to $75 million by the end of the biospherians' two-year mission, by various people's estimates. The silhouette of the Biosphere's jumbled domes and pyramids themselves silently told the stories of excess. The tall, phallic library tower above the Human Habitat, with its sweeping views of the surrounding desert, was the crowning monument to the project's costly grandeur, a tower on a castle.

According to its corporate mission, Biosphere 2 was theoretically, somehow, supposed to be making money. During construction, Ed Bass optimistically claimed in an interview, "It is a somewhat speculative and risky, but still very sound type of investment." He could not say exactly how Biosphere 2 would turn a profit, but had faith that profitable uses would emerge, he said.

> I've had the expectation—I think everyone else involved in that strategic level of the project has the expectation—that those sorts of discoveries, and the activities that will flow out of it then, will be there . . . It is principally a research and development type project, but it's not one that is expected will have to be continually funded and supported and so forth. At a point, once the investment is there, once the apparatus is there, it takes care of itself.[13]

Thus, just as the creators assumed that Biosphere 2 would ecologically "take care of itself," they also assumed it would somehow become financially self-sustaining. Maybe new inventions would come out of the project, they hoped. "We did in those days like to draw a parallel of how the space program developed technologies that really weren't conceived at the beginning but had great utility and great economic value," Bass told an audience of scientists at Biosphere 2 years later. NASA's celebrated "spin-off" technologies—earthly inventions based on technologies originally developed for space—ranged from solar panels to water purification systems, to lunar-imaging software that a cosmetics company used to analyze human skin in order to develop effective wrinkle creams.[14] Beginning in the 1970s, NASA put out glossy publications detailing around fifty new spin-off inventions every year. Still, Biosphere 2's fabled spin-offs were slow to appear as the Biosphere itself demanded most of the focus. Work began on the "Airtron," a commercial indoor air-purification system modeled after fans that blew air through the soil in Biosphere 2; and wastewater recycling technologies and environmental monitoring systems were other spin-off projects. However, these products remained in the development stages during the biospherians' mission.

Ed Bass had never wanted the spotlight, but as the costs climbed, the fate of Biosphere 2 would increasingly lie in his hands. He was not a typical billionaire, if such a thing existed. While his older brother Sid was building himself a $75 million mansion in Texas, Ed built a $150 million Biosphere but drove a little Toyota. In addition to funding Biosphere 2, Ed spent money and time sitting on the boards of the World Wildlife Fund and the Jane Goodall Institute. He called himself a "philecologist"—one who loves nature—rather than a "philanthropist," one who loves humanity. Ed had not set out to get rich. He and his brothers inherited their fortune from their great-uncle Sid Richardson, a Texas oilman. When the oldest brother, Sid Bass, invested the family fortune in Disney in the 1980s, the four Bass brothers suddenly found themselves transformed from multimillionaires, to multibillionaires. At first Ed's eccentricity quietly coexisted with his brothers' aims. The brothers poured millions into revitalizing their hometown of Fort Worth, an effort that included the Theater of All Possibilities' Caravan of Dreams performing arts center.

Even those closest to Ed Bass could only partially explain what was going on in Ed's mind concerning Biosphere 2's managers. For years Ed loyally supported John Allen and his other long-time friends who were running Biosphere 2. "He believed in them, and trusted them, and believed in the project . . . in general, he always gave them the benefit of the doubt in every single crisis," claimed Terrell Lamb, Ed's press representative. But, she said, "They had very specific ideas about how great they were at what they were doing, and they kept just telling Ed they needed more money."

Ed had already spent over $200 million as the biospherians' mission went on. Now he wondered how long that number would keep growing. In early 1993 Ed ordered Mission Control to employ a consulting firm to bring order to the finances. In came a powerful capitalist duo: Steve Bannon, a Los Angeles–based investment banker who specialized in corporate takeovers, and Martin Bowen, a quick-tempered banker who managed Ed Bass's money in Fort Worth. (Ed liked to call Martin "Bow-wow"; one Biosphere employee called him "a short, emotional Texan.") According to "Bow-wow," his and Bannon's goal was "to try to bring some financial discipline to the project" in order to curb "substantial expenditures in excess of budget." In reality, the battle for Biosphere 2 was on.

Under Bannon and Bowen's pressure, the Biosphere 2 team did begin working on new budget plans. CFO Marie presented a budget that she thought could lower Biosphere 2 operating expenses by a few million dollars per year, as Bass had requested. Desperate for some new source of funding, Bannon led the Biosphere staff in writing wishful marketing plans to sell anything that might possibly turn a profit. Biosphere 2 employees typed up plans to create an environmental education organization that could apply for grants, and plans to sell the software from Biosphere 2's data collection system. Meanwhile, Steve Bannon's Beverly Hills firm churned out grandiose proposals to create a "coherent global 'roll-out' strategy for artificial Biospheres and establish a system to exploit intellectual capital," using a "Disney-type control of its franchise strategy." The plan was to sell biospheres to governments and academic institutions around the world, as though everyone in the world might want their own after seeing the success of Biosphere 2. Bannon even started discussions with Veldon Simpson, the Las Vegas developer of megacasinos such as the Luxor, which featured a giant Egyptian pyramid and a replica of the Sphinx. The plan: to build "Biosphere 3," a casino and resort in Vegas.

Yet even as the Biosphere 2 management seemingly went along with cost-cutting measures and outlandish marketing schemes, the struggle for control of Biosphere 2 simmered under the surface. As Margret Augustine put it, "Various groups were vying for who wanted the Biosphere." Bass hired an accounting firm to audit the Biosphere company. John Allen insisted to me, "We were the marvel of any executive, capitalist or communist, at the low cost we were doing Biosphere 2 for, but it meant everything had to be closely watched."[15] However, that view was not shared by Bass's team. "The people at the Biosphere just treated huge amounts of money like it was petty cash," Steve Bannon claimed. "A bunch of plans were put up that they wouldn't meet . . . They continued to fail to meet certain objectives and goals and budgets and cash flows and all that." Bannon's firm made presentations to New York investment banks, hoping someone would give Biosphere 2 some more venture capital. Not surprisingly, they failed to find a new financial backer.

John Allen, despite the massive spending, was a shrewd businessman. He had set up dozens of interconnected corporations over the years of his Ecotechnics work, culminating in Space Biospheres Ventures. Though Bass had put up the money for Biosphere 2, his investment firm, Decisions Investment, was locked in a fifty-fifty joint venture with the rest of Decisions Team, the board members who had provided the sweat and intellectual capital to build Biosphere 2. Bass could not simply fire his partners, because they were in an equal partnership. The board members themselves received a nominal salary until the project was up and running in 1991; John Allen claimed only a $10,000 yearly salary plus housing. But the board members also part-owned Biosphere 2.

Though tensions were shaking every level of the Biosphere 2 organization, the biospherians did complete their mission. An eager crowd of five thousand spectators and two hundred news reporters flooded the lawn outside the Human Habitat to watch the biospherians make their grand "reentry to Biosphere 1." In a ceremony designed for the public and media's benefit, again, the biospherians donned their navy blue space suits—noticeably baggier now on their thin bodies. Walking in a line, they filed out of the airlock door, squinting just to stand in direct sunlight again for the first time in two years. Free at last, they strode in slow motion toward the stage, under a sky and in front of a crowd that both had never seemed so vast. Ed Bass turned from the microphone to smile and greet them: "Welcome home." Representatives from a local grocery store were on hand to present each biospherian with a complimentary bag of specially requested goodies that they had craved during their two years: chocolate, brie cheese, Roy's long-dreamt-of bottle of scotch. But the biospherians had become so attuned to the Biosphere 2 diet, lacking refined sugars, that even carrots had come to taste sweet to them, and the processed food actually made them feel sick. To those accustomed to eating food just harvested from their own fields, Biosphere 1's supermarket produce tasted old and stale.

"Not until Biosphere 2 has there been a structure in which to analyze the interaction of human beings with their total environment," Margret Augustine wrote in the "reentry" issue of the *Biosphere 2 Newsletter*. "And, to our pioneers, the first explorers into Biosphere 2, thank you for your selfless commitment and for reminding us that human intelligence and determination can accomplish great things when there is dedication to a cause larger than ourselves."[16] Meanwhile, the management tried to search the biospherians' belongings on their way out, and only relented when Taber threatened to tell the press.

Incredibly, the theater production went on. A new group of biospherians was deep into training for Mission Two. The new crew would number seven—five men, two women. Gaie, named scientific director of the five-month transition period, dove into preparing Biosphere 2 for its new inhabitants. She still felt intensely connected to Biosphere 2—so intensely that one night, when she could not sleep in her bed down the road from the Biosphere, she went back and slept in her Biosphere 2 apartment. She likened the experience to what explorers might feel like after living on other planets: "Some of the first space travelers to go to Mars and live there, they may never wish to go back, because once you adjust to that world, and you're in it, that's the response of the body you have: this is where I'm from now."

In the last month of Mission 1, more oxygen had been pumped into Biosphere 2 to make it easier for unacclimated scientists to breathe inside. Now teams of scientists entered through the airlock daily. Research consultants and Biosphere 2 staff remapped every plant in the wilderness, measured biomass changes, and closely analyzed the soil, water, and air in each biome. They introduced new insects, brought more plants into the wilderness to fill in gaps where others had died, installed bright high-pressure sodium lamps over the agricultural crops to make them grow faster, and planted new crops. Then, on March 6, 1994, the sequel opened: the new biospherian crew entered Biosphere 2 for Mission Two. Again the ritual speeches of a closure ceremony were performed, though to a smaller crowd of five hundred. Again the airlock doors were sealed. Press releases announced that some biospherians would stay inside Biosphere 2 for one full year, while others would come out after a few months to allow outside scientists to enter the enclosure as short-term biospherians.

Time was already running out on Space Biospheres Ventures, however. Finally Ed Bass got the law on his side, by going to his hometown back in Texas. He persuaded a U.S. District Court judge in Fort Worth that Decisions Team had mismanaged Biosphere 2. In a sworn affidavit, Ed testified that he believed that his old friends, faced with losing their life's work, could sabotage the Biosphere's atmosphere and its research. The court issued a temporary restraining order against key members of the management, including John Allen, Margret Augustine, Mark Nelson, and Marie Harding.

April Fool's Day 1994 was also Good Friday; it began as a quiet morning at Biosphere 2. John, Laser, and Gaie, who had become head of research, were in Japan seeking funding. Margret was away visiting family in Canada. "It's a slow news day," PR man Chris Helms told a friend on the phone. Then, at approximately ten in the morning, a fleet of vans and sheriff cars drove up to the gates of Biosphere 2. Out poured a small army of U.S. marshals holding guns, followed by a posse of businessmen in suits, a corporate battalion of investment bankers, accountants, PR people, and secretaries. They had come to take over. Off-duty police officers quickly secured the property, changed the locks on the doors and computer access codes, and established an armed command post. Though many of the top managers lived in apartments just downhill from the Biosphere, they were barred from their offices. The police guards used walkie-talkies to track the movements of longtime Ecotechnics members around the site as though they were suspected criminals. Bowen and Bannon convened an all-staff meeting, at which Bowen announced, "I am now responsible for overseeing this project."

Biosphere 2 was under occupation. The change in management regimes bore more likeness to a coup d'état than to a corporate takeover. The scene as described by Chris Bannon, Steve's younger brother and business associate, made it clear that Biosphere 2 was no longer the subject of a business dispute, but a battlefield: "I personally was tasked to be the first one out. We had to secure the gate out front, get out with the federal marshals, serve the security person at the gate, secure the gate, and then move into the other critical areas that had already been defined in order to ensure the safety and security of the Biosphere and its systems."

Steve Bannon had served as an officer in the U.S. Navy for seven years before becoming an investment banker. Now he became acting CEO of Biosphere 2. The takeover team saw the old guard of Space Biospheres Ventures as a dangerous enemy. They had planned the takeover for that day, while key leaders were out of town, said Chris Bannon, because "part of the team we wouldn't have to deal with initially, so we'd have less resistance. Marie was here, and obviously some of the top lieutenants were here . . . Bill Dempster was one of the die-hard loyalists—die-hard." Off-duty police continued to patrol the property day and night for fear that the "loyalists" might strike back.

The employees and residents of Biosphere 2 had varying reactions to the takeover. When Bass's publicity representative, Terrell Lamb, came into Chris Helms's office and told him, "We're here to take over," he fell into her arms sobbing in relief. When Martin Bowen's and Steve Bannon's faces first appeared on the biospherians' videoconferencing screen inside Biosphere 2 to tell them the news, it sounded so strange that the biospherians first assumed it was a practical joke or a bizarre training exercise to test their resolve. "We had no idea what was happening," biospherian Charlotte Godfrey said. "It was April Fool's Day, and here they are, saying, 'We've taken over the Biosphere, we've taken it back in the name of Ed Bass.'" The biospherians were given the option to leave, but chose to stay inside.

While John Allen stayed with friends in Japan, Gaie and Laser, who were not listed on the restraining order, flew home and returned to Biosphere 2 in the early hours of April 4, driving through the predawn desert by a back road. While the Mission Two biospherians slept inside, the pair smashed small glass safety panels to neutralize the Biosphere's air pressure, then threw open the airlock doors. They quickly left, then telephoned the biospherians, telling the crew that they now had the freedom to leave. Journalists excitedly declared "sabotage," but Gaie and Laser maintained that they were concerned about the biospherians' safety. Opening the doors, they argued, would make the crew feel more free to walk out of the compromised experiment. (However, the biospherians poked their heads out, closed the doors, and chose to go on with their mission.) It was an act of subversion by two of the creators of Biosphere 2. To Gaie and Laser, seeing bankers take over Biosphere 2 was like watching their world come under foreign occupation by an enemy. "We didn't know what to do. This was such a devastating thing to see happen to a biosphere that we loved so deeply . . . I still felt more attached at that time to Biosphere 2 than Earth," Gaie said. She was convinced that history was on the line, she told me, speaking urgently:

> I realized . . . that this was a historic decision, and that if I sat by and didn't do anything, that history would be changed forever, and that the future of closed systems would never be achieved, that long-term human inhabitation in space would be forever jeopardized, because as [former NASA astronaut] Joe Allen said, this was similar to a crew being up in a space mission and having Mission Control stormed below by cops.

"My only option was to open those doors," she said. "It was a pretty frightening decision to make . . . and then they ran after us with helicopters, and we were running for our lives . . . They sent helicopters after us, looking for us. We were walking back to our car, we were leaving, and we saw them. We heard from our friends that they were storming their apartments with guns . . . This was a project under siege. This was out of some really weird war movie."

Three days later, Gaie and Laser were apprehended by the Arizona police. "We were arrested without an arrest warrant, we were fired, our apartments were searched," Gaie said. "And then we were sued in the federal court of Texas by the billionaire bullies." So began a series of drawn-out court battles between Bass's representatives and the two ex-biospherians.

That June, three months after the takeover, after a final flurry of negotiations, the members of Decisions Team signed a confidential, "mutually satisfactory" agreement with Ed Bass. The terms of the agreement, such as any exchange of money involved, were secret, but it was not hard for an outsider to deduce some of the results. Afterward Bass would keep Biosphere 2, as well as the Caravan of Dreams in Fort Worth, the cattle ranch at Quanbun Downs in Australia, and the Hotel Vajra in Kathmandu. The rest of Decisions Team ended up still holding several of their old Ecotechnic projects: Savannah Systems in Australia, the Puerto Rico rainforest property, the London art gallery, and the farmhouse in southern France.

Humanity had once been kicked out of the Garden of Eden for aspiring to divine knowledge. And with their signatures at the bottom of the secret agreement, one by one the creators finally let go of their new world.

ACT IV

The former Intensive Agriculture
Biome awaits a new experiment.

PHOTO BY MARK GEIL.

THE RESET BUTTON

"A POET BANISHED FROM THE SITE HE DREAMED," HE CALLED HIMSELF.[1] JOHN Allen sought to make sense of the Biosphere 2 story in verse. He became his playwright alias Johnny Dolphin once again, a poet and wanderer of the world. But his mind never left Biosphere 2 behind. His new poems roared about "The Takeover" of his dream world by a greedy businessman named "Louis Cannon," working for a billionaire named "Mr. Trout."

For the builders, the takeover of Biosphere 2 was "sort of like a bomb blowing up and we all scattered, willy-nilly here and there," said Gaie. She and Laser ended up deep into legal battles with Ed Bass—lawsuits and countersuits, criminal and civil charges of sabotage over the pair's Biosphere 2 break-in, arguments over back wages, and accusations of libel and defamation. (Eventually the criminal charges against Gaie and Laser would be dropped, and they would settle the civil case with Bass out of court.)

Meanwhile John Allen and several of his remaining companions retreated to the hills of Southern California, in a friend's ranch house among avocado orchards. Marie Harding, John's former wife, moved back to quiet Synergia Ranch, where it had all begun. She started renting out the empty buildings to conferences and other visiting groups. Both of her parents died around that same time, but it was the loss of Biosphere 2, she said, that hurt the most.

In his poem "The Takeover," John Allen described the forces he saw closing in on Biosphere 2:

The bankers think an idea's a thing;
The bankers think a Biosphere's
A piece of engineering;

Bankers think money rules the world.
Courts agree, Cops agree,
Lawyers agree, Politicians agree,
Billionaires take a bow,
Sell-outs take a cut,
Reductionists cut out the total system,
Journalists and anchors agree,
Artists portray what gets funded.
Scientists research what they're granted.[2]

Parts of his poetic interpretation were not far off from reality. Even as seven new biospherians still ate, slept, and worked inside the glassed-in world, Biosphere 2 was entering a new era. The campus would continue to be a stage for varying characters acting out their visions of nature and the human place in it, but now a strikingly different vision would reign. Biosphere 2 was already many things to many people; now the structure became even more schizophrenic, as a posse of money-minded business managers set out to tame the beast, convinced that they could transform it from a luxuriant miniworld to an efficient and prestigious research machine.

Multiple interpretations of Biosphere 2—and of nature—now vied with each other. Steve Bannon, the new acting CEO, took an investment banker's view of Biosphere 2: "Let's get these biomes up to speed, let's get on with it," he told me was his attitude. Steve's younger brother Chris Bannon, working under him, recalled, "Our mission was to, one, button the place up and make it look and function like a business, as economically as possible." Their second mission: "find somebody to run this place," he said.

The Bannon brothers combed through record books and found the finances in chaos wherever they looked. Even some enterprises that were supposed to be sources of revenue were losing money. The visitors' center, though it was drawing over 200,000 tourists annually, was still losing a million dollars a year. "It was just a mess," said Steve Bannon. A former navy officer as well as a businessman, he took command. He ordered each office at the Biosphere to document its operations, then started restructuring the company and laying off old employees.

Biosphere 2 was not a new world in Bannon's eyes. Biosphere 2 was an ailing commercial enterprise.

Meanwhile, the seven new biospherians were still sealed up in their glass castle—but now they were unsure of their mission. They still imagined themselves to be living in a separate world, but the mandate for doing so seemed to be dissolving. Rodrigo Fernandez, a twenty-four-year-old biospherian from Mexico, said he felt like the Soviet cosmonaut who had been on the Mir space station when the USSR collapsed back on Earth—he was still drifting in space, but suddenly all the authority structures and sense of purpose back on the ground had become uncertain. Rodrigo wanted to stay in Biosphere 2, but he and his crewmates felt unclear about their mission: "I'll stay in here, but tell me why, what am I supposed to achieve, what's our objective? Give me an objective and I'll stay in, no problem, but don't just leave me here," he recalled thinking.

So the biospherians continued their daily work and tinkered along, waiting. Learning from their predecessors' struggles, they did enjoy some successes. Biospherian farmer Tilak Mahato, a new recruit from Nepal, coaxed much more food out of the Intensive Agriculture Biome than the first crew had. Tilak was humble and spoke softly, but he had a college degree in horticulture, and he knew plants intimately; he was, in the words of one of his fellow biospherians, "a wizard of tropical agriculture." He also had the benefit of new cultivars that the agricultural consultants had found for Biosphere 2's unique conditions after the first mission's difficulties. During Mission One, the biospherian diet had dwindled to only a few reliable crops; the biospherians had stopped planting any crops that presented problems, fearing that if they did not grow enough food, they would starve. Now Tilak experimented with new crops, assured by his new bosses' promise that if the crops failed, they would import food rather than let the biospherians starve. The first crew had planted corn but gave up on it as a crop early on because it did not set seed well. Tilak tried again, pollinating each stalk by hand, since corn in Biosphere 2 had no wind to spread its pollen. He was rewarded with abundant harvests, which his companions ate gratefully.

The biospherians experimented in other areas as well. Their new bosses knew next to nothing about ecology, so the crew had more room to play around. Charlotte Godfrey, the biospherian in charge of the wilderness areas, recalled that under the old Mission Control, she got yelled at for pruning the rainforest without instructions, but now she felt free to do as she pleased. "It was really fun, because it was like going into a grandma's attic—hey, what does this machine do? Try it out," she said. When Charlotte took on the rainforest, the soils had grown salty and hard, unhealthy for plants. "There were pools of water on the soil, this hard clay soil. You could smell the anaerobic bacteria, because it would just lay there. And there was gypsum salt, just layers of gypsum salt in the soil," she said. During Mission One, the biospherians had made regularly scheduled rains, with the goal of stimulating plants to grow and suck up CO_2—but the rain had come from salty runoff water that left salt deposits behind in the soil. Charlotte decided to see what she could change to boost the health of the rainforest's soils and plants. First she gave the rainforest a heavy shower of precious clean water, then subjected it to drought for several weeks to see how the plants would react to experiencing different seasons.

However, it soon became clear that biospherians growing corn and tinkering with the rains did not excite the new management. "They didn't want to spend anything," said Charlotte. "The first management put such a huge emphasis on the living systems, they were such a priority . . . and little by little, I saw that priority fall by the wayside." Gradually, the rationale for the human experiment eroded. Steve Bannon's view was straightforward, he told me: "What was being gained by locking these people up for a year?"

The biospherians began to wonder the same thing. The crew began to shake up. Twelve days after the armed takeover of Biosphere 2, biospherian Norberto Alvarez-Romo, the former Mission Control director, chose to walk out of the Biosphere. Norberto was married to Margret Augustine, who was legally banned from the premises; he found himself in an impossible situation. Bernd Zabel happily stepped in to take his place, having waited years for the opportunity. A long-time synergia member, Bernd had been appointed captain of the first biospherian mission, only to be fired at the last minute before closure. Now he got his chance.

The other new biospherians were a mix of old and new characters. Two other crew members were long-time Institute of Ecotechnics associates: Pascale Maslin, who had lived at the Ecotechnics farm chateau in southern France and sailed around the globe on the *Heraclitus* in the 1980s, and John Druitt, who had spent years living at the Puerto Rico rainforest project. But the other four biospherians were newcomers. For the first mission, fourteen candidates had eagerly competed to be locked inside Biosphere 2. Many of those aspiring biospherians had lived and worked together at Institute of Ecotechnics projects for over a decade. But most of them had drifted away from Biosphere 2 by the time Mission Two was getting under way. The management had trouble even recruiting enough biospherians for the second crew. Charlotte Godfrey, the biospherian in charge of terrestrial systems, was only twenty-two years old when she entered Biosphere 2 in 1994. A Tucson native, she had started work at Biosphere 2 two years earlier, as a greenhouse caretaker, after she saw an ad for the position in a local newspaper. When Charlotte began her work on the Biosphere 2 campus, the Mission One biospherians were already locked in, but among the long-time staff outside, "there was a mass exodus; a whole bunch of people quit," she recalled. Still, Charlotte enjoyed the greenhouse work, she said; the campus still had "that old-fashioned air of doing something amazing." The other new biospherians each came to Biosphere 2 through their own coincidences. Tilak Mahato, the new biospherian farmer, met John Allen when John visited an eco-tourist resort where Tilak was working in Nepal. Rodrigo Fernandez, a Mexican who had just gotten his engineering degree, met Norberto Alvarez-Romo through his academic advisor. Graduate student Matt Finn, the biospherian in charge of marine systems, came to do his dissertation research in Biosphere 2's mangrove marsh, but after three months he had collected the data he wanted, so he left, to be replaced by Matt Smith, another young marine scientist who had worked on Biosphere 2.

Slowly, as the biospherians changed places, the concept of Biosphere 2 as a separate new world was dissolving. Visiting scientists began to join the biospherians for stays ranging from a day to two weeks. On one day in May 1994, thirty-one scientists and technicians entered for the day to check out the wilderness. The core biospherian crew still tried to hang onto the old dramatic sense that they had left the Earth behind, but it was difficult. Rodrigo still tried to imagine Biosphere 2 as "a space station, where you have a base crew that just keeps the place operating, and you have research crews that come in and do their stuff and then they go back," he said. But the analogy had its limits. The biospherians enjoyed the company of a few long-term guests, but they became annoyed at the increasing flow of maintenance crews and one-day visitors—particularly when a visiting scientist burned the biospherians' coveted peanut harvest while roasting it in the oven. It began to seem pointless that the biospherians themselves never got to leave. As Charlotte said, "After a while it got to be like I wanted to be able to go out and have a steak, because it didn't make any sense that we had to stay in there twenty-four/seven."

At last Biosphere 2's atmosphere made its own decision about the biospherians' fate. Ever since the first closure, levels of nitrous oxide (N_2O)—dentists' "laughing gas"—had been slowly building up inside Biosphere 2. The gas is one by-product of soil microbes' respiration. In Biosphere 1, N_2O breaks down under ultraviolet sun rays in the upper atmosphere. But UV light waves could not penetrate Biosphere 2's glass walls. By Mission Two, N_2O levels inside Biosphere 2 exceeded the safety limits set by the U.S. government. (The regulations had been set to protect dental assistants who administered the gas to patients, because prolonged exposure could damage the human nervous system.) Finally, worried about their health, the biospherians asked to come out. They stepped out of the airlock on September 17, 1994, in a small "reentry ceremony" barely six months after their mission's start, ending the last human inhabitation of Biosphere 2.

The removal of the biospherians was part of a larger process of transformation. The whole idea of Biosphere 2 was changing, reflecting a different vision of nature and the human place in it. In its new directors' eyes, the wild living world was a mess; it needed to be transformed into a clean machine. With the biospherians gone, the new managers flushed the entire Biosphere with fresh air and water from outside—permanently washing away the original idea of a closed self-sustaining system in the process. Chris Bannon described Biosphere 2 as though it were a machine that could simply be purged, cleaned out, and started over:

> Flush the Biosphere out, clean it out; let's set the meter back to zero. So we did a press release and developed a plan in the press release calling it the "big flush," and we went in and did a huge massive clean-up on the Biosphere. There were eight hundred drums of by-product from the CO_2 scrubber, there were these bales of biomass that were just stacked in the Biosphere, the roaches, there was moist stuff sitting in front of air handlers with spores and mold counts . . . It just needed to be cleaned out

and recalibrated and then people could go in and find out what it could be used for.

A team of workers, wearing gas masks to protect themselves from accumulated mold spores, set to work hauling the bales of dried plant matter—carbon that the biospherians had harvested from the atmosphere—out of the basement, weighing it, and exporting it from Biosphere 2. Cleaning out the whole structure, and giving it fresh air and water, would be "essentially hitting the reset button" for Biosphere 2, Steve Bannon told a reporter.[3]

But in other ways, Biosphere 2's new rulers seemed frightened that the ghosts of the past could not be cleaned out so easily. They showed a profound unease toward any reminders of the project's founders. Bannon and Bowen purged the staff of suspected loyalists to the old regime. Long-time employees were not allowed to call each other by their old nicknames on the job. William Dempster, the engineer who knew Biosphere 2's machines better than anyone else, got to keep working there at first because he was indispensable, but later he was fired because the new managers feared he was a spy for John Allen.

Martin Bowen, the fiery Texan banker, watched closely to make sure the cultural takeover was complete. He ordered maintenance workers to destroy any physical artifacts of Biosphere 2's founders. On his instructions, workmen demolished the Native American sweat lodge on the hill behind Biosphere 2, and dismantled Tibetan chortens and rock sculptures by John Allen and Laser. Metal robot sculptures had peppered the Biosphere campus. A Southwestern artist had welded the sculptures using recycled parts from the Los Alamos nuclear laboratory; to the founders of Biosphere 2, they were a whimsical representation of how to create new art out of old industrial destruction. Now, imprisoned in a storeroom on site behind bars, the sculpture collection became a visual metaphor for the new managers' attitude toward Biosphere 2's past: they wanted to lock it up, but it still sat there, lingering.

Still suspicious of any lurking spirits from the past, acting CEO Steve Bannon even called for ritual purification to cleanse the old energies of Biosphere 2. Bannon quietly asked Steve "Bear" Pitts, a staff scientist descended from the Cherokee medicine tradition, to ceremonially cleanse the place. "There were people who would not work inside the Biosphere, maintenance crew, because of their fear of what might be there," recalled Bear, a quirky computer modeler with long gray hair. "And so I was basically told by Steve Bannon one day that there would be nobody around, and for me to do my work." Bear burned sage in Linda Leigh's old office at her request; he then did the same in the ranch house across from the Biosphere, where the Theater of All Possibilities had played, and later in the house where Margret Augustine had lived, where Taber and Jane would soon have their wedding. Finally Bear moved on to purify Biosphere 2 itself. He moved around the structure with bells and pipes for six hours, interacting with all sorts of spirit energies. Bear was not surprised, he said, at the new business managers' request for purification: "If the spirits exist, then you don't want to piss them off, you want to enlist their assistance. And you want to give them an opportunity to either stick around or go away." Even the new corporate bosses of Biosphere 2 realized what intense energy had moved through that place.

The ritual cleansing of Biosphere 2 was not just physical and spiritual; it was also taking place on a professional level. At last, another set of priests was called in to sanctify Biosphere 2: the high priests of science. Steve Bannon was a corporate turnaround specialist, and Ed Bass had given him a clear mission: get someone else to take the project forward. To do that, Steve Bannon believed, "We need to get the best minds in the world here." The assumption was not so different from the driving idea behind John Allen's "synergetic civilization": create the best possible world by bringing together the best minds. But he would go about it in a different way.

Just as the resonance of the word *science* had been used to tear down Biosphere 2's reputation, now that word would be used to build it up again. Big-name men flew in from big-name institutions to weigh in on Biosphere 2's future. In July 1994, Michael Crow, the vice provost of Columbia University, flew out to Arizona to chat with the Bannons. Crow was a big proponent of global environmental studies at the university, and Wally Broecker, the Columbia scientist who had previously consulted on the Biosphere's oxygen problem, had been keeping his eyes on Biosphere 2 ever since. "I considered it a tragedy that this incredible facility was being used to accomplish what I viewed to be a frivolous goal," Broecker later wrote of the biospherians' human experiment.[4] Now that the project's founders were gone, he was eager to turn Biosphere 2 into his own project on climate change science. In August 1994, Columbia University signed on to manage the Biosphere 2 research program. University administrators wanted the school to become a world leader on environmental issues, and they saw Biosphere 2 as a shining opportunity to conduct large-scale climate change experiments, said Michael Crow. "There was lots of observational data related to global climate change, global warming, but there was no experimental data; there was no data about what might or might not happen, derivative of experiments of any size. So we began to look at the Biosphere as a facility that would allow us to have experiments." Bruno Marino, an isotope geochemist from Harvard, became Biosphere 2's new scientific director.

Columbia's leaders started to play more of a hand in the management, and staff shake-ups at the Biosphere continued. "Some could do the job, some couldn't, some people quit, some people were fired," Michael Crow recalled, but he saw these as "normal start-up problems." Biosphere 2 would now get down to "good, basic, objective science," new science director Bruno Marino announced in a news article in the pages of *Science*, the prestigious professional journal. "We need to show that we're credible, that we really are going to do good science," he said.[5] NASA's Gerald Soffen, a disgruntled past member of the Biosphere's Scientific Advisory Committee, had become an outspoken critic of John Allen; months earlier, also in the pages of *Science*, Soffen had called Biosphere 2 "a mini-supercollider that's being wasted."[6] Now he changed his tune: "This is great news . . . It sounds like science is finally taking over." All this fanfare came in spite of the fact that the *content* of Marino's promised "good science" was at that point almost totally unarticulated, and hardly more specific than the vague ecological goals that Biosphere 2 spokespeople had been announcing for years. The *Science* article only mentioned that "researchers from fields such as ecology, plant biology, and soil physics are being commissioned to visit Biosphere and determine what science could be done there—for example, studies of how plant growth is affected by high concentrations of carbon dioxide." The purpose of Biosphere 2 was still undefined; as Wally Broecker, Columbia's star geochemist, put it baldly, "We inherited a $200 million structure, and we have to figure out what to do with it."

In fact, the hypothesis driving Biosphere 2 became a social one: big names would bless Biosphere 2 with "good science," everyone seemed to agree. The Biosphere's new research directors commissioned proposals from scientists at major institutions all over the country: Columbia, Harvard, Yale, MIT, UC Berkeley, Stanford, and other universities; the Santa Fe Institute; the U.S. Department of Agriculture; and various government laboratories. The resulting collection of white papers, an August 1994 press release announced, would "determine the optimum science program," because the authors were all "prominent scientists in their respective fields." Steve Bannon wanted to pull in "the Wally Broeckers and the world-class guys, the Nobel Prize guys that you must have at a place like Biosphere 2 to make it credible," he said. Wally Broecker himself agreed: "Only if we can build cooperative efforts involving high-profile people at leading institutions (and also their students and postdocs) is there a chance that we can reach our goal of establishing the Biosphere 2 campus as a world-recognized center for biotic research."[7] Wally was known for his sharp tongue and difficult temper, but as Steve Bannon put it, "These scientists are like prima donnas, but it's worth it if you want to work with the best."

When the consulting scientists convened at Biosphere 2 to present their proposals in December 1994, the scene was different from the early dreaming days of the planning conferences for Biosphere 2. In the early years, recalled Tony Burgess, "We were an eclectic group of naturalists, we were designers, a lot more giggling and playful," but now the tone was more solemn: "There was a lot of serious discussion about what was good science, what was bad science." The scientists themselves were a different breed, who all shared one important characteristic, Tony said. "They were all people very much in the center of the scientific establishment, all of them stars or rising stars in getting big grants."

The rising stars took turns offering their proposals for Biosphere 2, landing all over the map of the life sciences. Former Biosphere 2 science director Jack Corliss, along with the Santa Fe Institute's Chris Langton, suggested using Biosphere 2 to develop computer modeling techniques for studying ecocomplexity. Bruno Marino wanted to track isotopes of various gases, particularly nitrous oxide, methane, carbon dioxide, and oxygen, as they moved through the Biosphere. Biologist Peter Vitousek of Stanford suggested studying the interrelationships of the different "wilderness" areas to look at the importance of ecosystem diversity, though as for who should undertake such an experiment, he only offered, "Not me." A U.S. Department of Agriculture group wanted to plant wheat in the Intensive Agriculture Biome and compare it with USDA's own experiments on the effects of carbon dioxide on wheat plants. Many of the proposals centered on the topic of carbon dioxide, approached from a variety of angles: how would increased carbon dioxide affect photosynthesis, plants' water use, soil processes, crop growth? Following bureaucratic academic process, five committees emerged from the visioning conference, pledging to further explore research possibilities in five areas: the carbon cycle, elemental and trace gas cycling, biodiversity, sustainable agriculture, and engineering.

But in the quest to redefine Biosphere 2 as a research tool, the old image of a self-organizing, evolving living system seemed to have fallen by the wayside. One white paper author did argue, "To the 20th century scientist, used to the research mode of grabbing one component, jerking it, and studying the response, a useful caution occurs. Much of the value of Biosphere 2 may in fact be obtained by humbly watching and characterizing the damn thing for a long period of time . . . Darwin would have loved this place."[8] However, Darwin was not in charge of Biosphere 2. And its new leaders, desperate to prove themselves, did not have a long time for "humbly watching."

Despite the lack of consensus over how to use Biosphere 2, its new stewards began claiming victory anyway; at last Biosphere 2 had been blessed by the high authorities of science. Michael Crow, a champion of Columbia's involvement in the project, proclaimed in a press release, "The most important thing to come out of the meeting was that a group of world-class scientists got together and decided that the Biosphere 2 facility is an exceptional laboratory for addressing critical questions relative to the future of Earth and its environment. That has not been validated before."

In fact, it was a new type of scientific theater. Biosphere 2 was becoming an "exceptional laboratory" not through scientific exploration, but through the institutional appearance of academic discourse. Michael Crow admitted this openly, years later. "We were having significant difficulty trying to determine the scientific agenda," he told me, then corrected himself: "We knew what we wanted to do, but we also wanted the endorsement of the broader scientific community. So we began bringing in large numbers of external advisory groups." Biosphere 2 had once been attacked as "unscientific," largely because of the personalities involved in the project; now, in the project's reinvention, the definition of "science" was still tied less to what was being said than to who was doing the talking.

Steve Bannon continued to venture into academic politics, discussing with administrators at Harvard, Columbia, and Yale to try to get someone to commit to Biosphere 2 for the long term. In the end, Columbia agreed; Columbia scientists and administrators were already essentially running the research program. Biosphere 2 would also fit neatly into the university's new Earth Institute, an interdisciplinary network of environmental research centers. In November 1995, after months of negotiation, Columbia's leaders signed a deal to manage Biosphere 2 for five years, provided that Ed Bass put up some of the funding. At the end of the five years, the university would have three options: freely walk away, renew its contract, or buy Biosphere 2 from Bass for a mere $1 million—pocket change compared to the original $150 million construction cost.

The Columbia era at Biosphere 2 officially began on New Year's Day 1996. At the front gate, out on the highway among the cacti, a powder blue flag bearing the school's insignia rose into the desert wind. Below it sat a huge boulder with block letters freshly carved into it: "BIOSPHERE 2 CENTER / COLUMBIA UNIVERSITY." The Ivy League school's name meant that Biosphere 2 had finally achieved respectability, Martin Bowen told me. "They don't associate with bad projects. They don't associate with bad deals." Bill Harris, later president of Columbia's Biosphere 2, made the same point to a reporter: "What impressed me was that Columbia University had the courage . . . that they would put their name, which is their most valuable asset, on this property."[9]

The press took a similar tone. As in the past, the nation's reporters all seemed to change their mind about Biosphere 2 at exactly the same time, again showing the power of popular media to create their own reality by the sheer force of repetition—and again showcasing the common assumption that "good science" could be neatly and simply defined. The *Arizona Daily Star*, one of the state's main newspapers, had bashed Biosphere 2 for years; now its lead editorial argued, "By enlisting the aid of Columbia and Harvard universities in nurturing legitimate research, the project has taken an overdue step toward returning to the fold of respectable science." The article suggested that "respectable names promise respectable science."[10]

Likewise, "[a]fter years of scorn for emphasizing flashy survival experiments over basic research, the controversial Biosphere 2 project appears to be on the verge of scientific legitimacy," the *Christian Science Monitor* announced in early 1995.[11] Reporters were surprisingly quickly convinced by the Columbia name. An Associated Press story, printed in newspapers all over the country, announced that now Ed Bass would "put aside Biosphere's image as a blend of fuzzy science and New Age philosophy and legitimize the glass dome as a research tool accepted by mainstream scientists."[12] "Biosphere entering brave new world of hard science?" wondered a headline in the *Atlanta Journal Constitution*, continuing a tradition of mixed metaphors in the project's press coverage.[13] Indeed, there seemed to be a general agreement that the new science would be not "fuzzy" but "hard," and that this was a good thing—a decidedly yin-yang imagery associating "real science" with cold, hard masculinity. ABC News's website sealed the verdict: "Biosphere 2 a joke no longer."[14]

But perhaps Biosphere 2's speedy image-reversal—before any "hard science" had really taken place, or even been planned—should not be seen as so surprising. After all, the project was only reflecting the usual authority and awe that had long been associated with the elite scientific establishment. Scientific "truth" is supposedly based on physical facts, but those facts become widely accepted based on *trust*, as Steven Shapin has argued in *A Social History of Truth*. Life in modern civilization is complicated, and individuals cannot verify all knowledge for themselves; instead, they must trust others, such as scientists, to speak for nature. But whom to trust, in a complex modern world where it is impossible to know everyone else? In such a world, individuals place their trust by following cues such as academic degrees, institutional affiliations, job titles, and awards—"systems of expertise," in Shapin's terminology.[15] As a consequence, a narrow sector of society (dominated by white males) become the experts on nature. Sociologist Robert Merton has called this process "the Matthew effect in science," after the biblical dictum "for unto every one that hath shall be given": those who already possess the most scientific honors (such as a Nobel Prize or a top university professorship) are the ones who receive even more honors. In a social study of scientific prestige, Merton found that scientists with well-established reputations get disproportionately more praise, awards, and research grants than lesser-known scientists who do comparable work.[16]

Of course, despite all the fanfare, Columbia would still have to prove its promises of "good science" to skeptical colleagues in the scientific community. The glass world was still the butt of jokes at some academic conferences. Heidi Barnett, an enthusiastic young marine biologist, got her first science job as a research manager in Biosphere 2's ocean; her former university advisor warned her, however, that "if you want a career in science, that's academic suicide; you'll never get a job in science again," Heidi recalled.

Her new Biosphere 2 bosses felt the same worry: were they about to commit academic suicide? They seemed desperate to crank out anything that could be considered "good science," Heidi said. In Columbia's first years on site, "All the progression of the science was external pressure from Columbia to get something *done*—get something done *now*, I don't care what it is, get something done, make it good, get it published in *Science* and *Nature*." She recalled, "When Columbia got involved, I remember being in a site-wide meeting where [Columbia vice provost] Mike Crow, giving a series of baseball analogies, told us that we all needed to hit home runs, that we had to do 'good science,' that we had to do a kind of science that would be published in *Science* or *Nature*, we had to be in top journals, and we had to make a splash."

Power to determine research priorities was given to "hand-chosen people, Broecker and a few others, the money-makers, who were going to get in there and do quick and dirty 'good science'—what is the system ideally suited for and what can we get out—*fast*," Heidi said. Despite all the research proposals floating about, the final research program would largely be shaped by the interests of a small group of key men, most of whom were not even stationed at Biosphere 2. Those included Wally Broecker, Chris Langdon, and Taro Takahashi from Columbia, with major involvement from marine scientist Marlin Atkinson of the University of Hawai'i in the Biosphere 2 ocean, Joe Berry of the Carnegie Institution at Stanford in terrestrial areas, Kevin Griffin of Columbia in the Intensive Agriculture Biome, and Guanghui Lin, an on-site researcher at Biosphere 2, in the rainforest. "The evolution of current science is largely based on their minds," said Adrian Southern, who worked as a research manager in the Biosphere 2 rainforest under Columbia. "They were given the objective to come up with wowee-zowee science, something that would be front page."

THE NEW NEW WORLD

"WELCOME TO THE BIOSPHERE 2 CAMPUS OF COLUMBIA UNIVERSITY," the first sign announces. Visitors, clutching their $13 tickets, step through a brick archway. Trees and flagpoles flying university emblems line the path, alongside a gently sloping lawn. The walkway leads to a visitors' center, where large display panels on the walls explain "Biosphere 1: The Earth," "Biosphere 2," and lastly, "Columbia University, Biosphere 2 and You." This last sign is peppered with pictures: stone-faced scientists sawing apart tree rings; others scrutinizing glass bottles in a lab; and Columbia geochemist Wally Broecker, smiling in a sweater vest, his hands folded on top of a globe.

The glass pyramids of Biosphere 2 itself glint in the distance at the end of the campus's little valley. On their way there, tourists wander on asphalt paths through a ghost town of research greenhouses and gift shops, oddly silent in the sharp Arizona sunshine. They stop to duck into a little darkened theater where a video screen endlessly repeats a fourteen-minute movie entitled *Ahead of Its Time*. The film narrates the Biosphere 2 story through interviews with scientists, from desert designer Tony Burgess to Columbia's Wally Broecker—with certain key characters left out. If any biospherian ghosts linger, they are mostly invisible. (As biospherian Dr. Roy commented, "If you look at the signs around the Biosphere, you'd think that it just grew there and Columbia found it.") As tourists pad around the purple wool carpets of the quiet Human Habitat, they see a few blown-up pictures on the white walls: Roy taking blood samples; Taber looking studious in his white lab coat; Taber, Jane, and Linda laughing merrily over a cup of tea. (Again, it is no accident which four biospherians from Mission One, along with their friend John Allen, are conspicuously absent.) The biospherians' apartments have been converted into offices and are off-limits to visitors, except for one room, where an orange jumpsuit hangs on the doorway for show.

Back outside, the campus restaurant menu features the "Bowen Burger," half a pound of ground beef named after Ed Bass's head banker. In the Desert Digs Gift Shop, Rainforest Gift Shop, and Coral Reef Gift Shop, the inscriptions on pencils, refrigerator magnets, shot glasses, T-shirts, golf balls, and miniature souvenir spoons all announce whose Biosphere this is, lest anyone should forget: "Columbia University's Biosphere 2 Center."

Columbia's Biosphere 2 Center was its own new kind of world. Within the space of a few years, the "planet-in-a-bottle" evolved from Edenic survival camp to staid climate change laboratory. The ecosystems themselves had transformed. In the old Intensive Agriculture Biome, where biospherians once grew their food, now hung plastic curtains, dividing the space into three smaller rectangular enclosures; in each plot, spindly cottonwood trees grew out of bare dirt in neatly spaced rows, each plant identical to the next. Meanwhile, in the wilderness area, another huge plastic curtain hung from floor to ceiling, segregating the rainforest from the rest of the Biosphere. On the other side of the curtain, the savanna and desert sat neglected, dry and overgrown with grasses. The ocean lay placid at the base of the savanna's concrete cliffs. But the old corals had mostly died off, and researchers were paying more attention to tiny nubbins of new corals sitting in metal cages.

The changes were more than just physical. Biosphere 2 was shifting its identity; it was no longer a miniature world, but a laboratory, a giant test tube. William C. Harris, a former administrator from the National Science Foundation, became president. He and his bosses back at Columbia had big ambitions for Biosphere 2. University officials were putting together Columbia's Earth Institute, a collection of research centers in the physical, biological, and social sciences that aspired to be, according to the Institute's web page, "the world's leading academic center for the integrated study of Earth, the environment, and society."[1] In the wake of the Johannesburg World Summit on Sustainable Development in 1992, "sustainability" was not just for environmentalists any more; it had become a buzzword for major institutions seeking a role in the environmental future, said Michael Crow, the ambitious young Columbia vice provost who was a major force behind the Earth Institute. To Crow, facilities like Biosphere 2 were "physical and intellectual assets" for the university. Crow wanted Columbia to become an environmental power player, "focusing on using science to better sustain the planet," he said. Biosphere 2, and the university's Earth Institute in general, were manifestations of a broader process: traditional mainstream institutions, by the late 1990s, were finally taking on the environmentalist dream as their own—but they were also remaking that dream in their own image.

Columbia leaders continued with the quest to leave Biosphere 2's past behind. Chief of Staff Chris Bannon kept up his brother Steve's work of officially discrediting the Biosphere's founders. Chris photocopied an old news article calling John Allen a cult leader, and put it in all the Biosphere tour guides' mailboxes. Because of Bannon's attitude toward the past, several Columbia employees were afraid they would get in trouble for participating in interviews for this book. Some would only talk with me off campus or at least away from their offices, and one refused to use his Biosphere 2 email address to correspond with me.

Of course, there was an irony in Columbia's efforts to abandon the past of Biosphere 2: the facility's new stewards nurtured the same basic fantasy as the men and women who had created the place. These new Biosphere managers too were dreaming that they could somehow use a glass house in the desert to alter the fate of life on Earth. Columbia leaders might have scoffed at the grand space-colony plans of Biosphere 2's creators, who declared their intention to become "a creative collaborator with biospheres, rather than a parasite weakening the host."[2] But the university leaders' aims, cloaked in less colorful language, were equally grand, as reflected in new college semester courses taught at the Biosphere 2 campus, with names like "Planetary Management Laboratory and Seminar"—as though the whole planet were something to be managed in a lab.

The revised utopian dream of Biosphere 2 did contrast in some ways with the project's early visions. Biosphere 2 still harbored planet-sized dreams, but Columbia's leaders believed in a different path toward global change. As Michael Crow put it, the goal was "to move the Biosphere away from its emphasis on long-term sustainability survival missions, to broader scientific questions related to climate change and global sustainability." In the view of Crow and his colleagues, people did not need to build their own alternative miniworlds of artist-adventurer-scientists; instead the world as a whole would be alright, they seemed to imply, if people followed the instructions of wise scientists. A brochure for Biosphere 2 Center summarized this view:

> How will global warming affect our water? Our food? Our entire planet?
>
> During the course of atmospheric change, what will happen to various species—will they adapt? Compete? Die out altogether?
>
> As Earth's population grows, how will the planet support our needs for food, electricity, transportation, and medicine?
>
> What can human beings do today to protect the only home we know?
>
> At Biosphere 2 Center, we're asking questions like these—and finding answers.

"Peering into the future to see what our environment will be like 25, 50, even 100 years from now: That's the job of scientists at Biosphere 2 Center," the brochure read. Scientists, in this official view, had become seers who could foretell the future of the world. "The Window Into Our Future™" proclaimed plastic souvenir soda cups given out at the Biosphere 2 snack bar.

The Biosphere 2 structure now embodied a strikingly different view of humanity's proper place in the natural world. The scientists running Biosphere 2 were interested in the environmental future, but they were not interested in learning to live together and grow their food; they were interested in using their rational minds, from a safe distance, to control and probe the Biosphere for information. Biosphere 2 would still be used to model the Earth, but now in very different ways. Instead of trying to model utopia, Biosphere 2 would actually model *dystopia*—a future plagued by high carbon dioxide levels. And in subtler ways, Biosphere 2 would unintentionally continue to model human-nature relations in the wider world of Biosphere 1. For the glass earthbound spaceship was still a stage, where various power players—academics, corporations, and even the U.S. government—would act out their own visions of the environmental future.

The range of experiments proposed in the early days of Columbia's management reflected the deep shift in Biosphere 2's identity. Some scientists wanted to rip out the desert to create multiple mini-rainforests for side-by-side comparisons. Others proposed fumigating the biomes with pesticides and watching how they responded—darkly nicknamed the "Silent Spring experiment." Still others suggested burning down half the rainforest and watching it regenerate, in an experiment dubbed the "scorched earth" scenario. These more outlandish experiments were never attempted, but the fact that they could even be *discussed* at Biosphere 2 showed how much the place's culture had moved away from visions of a harmonious, Edenic wilderness and toward a view of nature-as-test-tube.

In the end, the research vision that won out prevailed partly because of the structure of scientific knowledge, partly because of the structure of scientific funding—and entirely because of the close linkages between the two. Columbia's leaders knew they had to get the buy-in—intellectual and financial—of the broader scientific community. They were keenly aware, as Tony Burgess put it, that "until Biosphere 2 itself shows that it can do valid, widely recognized, relevant scientific research, it's not going to get any federal money." The new Biosphere 2 therefore had to address top scientific and social priorities. And so one of the biospherians' biggest problems—rising carbon dioxide levels—now became the focus of research: how would various creatures and ecological communities adapt to increasing carbon dioxide levels in the world's atmosphere? As international diplomats were negotiating climate change treaties in the late 1990s and debating how much nations would have to reduce their carbon dioxide emissions, it was a timely topic.

Under academic politics, certain ecosystems in Biosphere 2 became privileged over others for the new research. Rainforests and coral reefs, sites of exotic appeal, became the focal points. Scientists began raising CO_2 levels in the Biosphere 2 rainforest to test how much CO_2 tropical rainforests could soak up on planet Earth. In the ocean, marine researchers tested how corals would react to changes in ocean chemistry under a high-CO_2 atmosphere. Meanwhile, the Biosphere 2 grassland and desert languished, largely abandoned. Biosphere 2 itself sat in the middle of the Sonoran Desert, one of the most biodiverse deserts in the world, an environment threatened by the exploding development of water-guzzling suburbs. But the desert was not a research priority, reflected Tony Burgess, who still worked for Columbia, sadly watching his carefully created desert become further overgrown with weeds: "It had nothing politically interesting to the United States." Eventually, thirsty for money, Columbia's managers would install a wide boardwalk path through the Biosphere desert and savanna, in order to herd tourists through for an extra $20 apiece.

Changes in the biospherians' former farm area similarly reflected power politics outside. In the Intensive Agriculture Biome, biospherians had once tended dense plots of their food crops. Now the fields became the Agro-Forestry Biome, a monoculture of skinny, cloned cottonwood trees growing in naked soil. Before, the field had represented a careful effort to feed people on a small-scale, intensive farm without using polluting chemicals—arguably an important area of research for the world's future. But such a program did not fit with Columbia's hunger for scientific funding. In contrast, the cottonwood experiments, using tree plantations to suck up carbon dioxide, were potentially attractive to powerful funders, including the U.S. government. Federal policymakers hoped they might one day grow such plantations to meet international climate change agreements. Fast-growing trees, turning atmospheric carbon into wood, might be a way for America to counteract its huge CO_2 output without having to cut back on the nation's addiction to fossil fuels. Such carbon sequestration experiments were also attractive, of course, to timber companies, such as Westvaco, the big Southeastern timber corporation that proposed the Biosphere 2 cottonwood experiment and donated the baby tree clones to start it.

To its founders, Biosphere 2 itself had been a living creature—"the ninth Biospherian," as Roy called her in one of his scientific publications.[3] Now the former biospherians disparagingly watched Columbia's progress from afar. Mark Nelson bemoaned the transition from the biospherians' "tender loving care" to "a paradigm change to scientists in white lab coats coming in from nine to five." But to Columbians like geochemist Wally Broecker, treating the miniature wilderness as sacred or lovable made little objective sense. To Broecker, Biosphere 2 was just an experimental device, he said. "They had built up this mythology as if it were a very delicate thing, as if you had to handle it just right," he complained of the biospherians. And though Broecker was stationed back in New York City, his word held influence at Biosphere 2. He was famous for his career studying the global climate and had won the National Science Medal. Tony Burgess, one of the last remaining staff from the early days of Biosphere 2, recalled what happened during Columbia's first season, when Tony was charged with keeping the Biosphere's wilderness healthy. Wally Broecker told Tony to switch the rainforest from overhead showers to a ground-based drip irrigation system. Tony protested that this was a bad idea ecologically; it could cause salt buildup in the soils and would not water the epiphytic plants hanging in the rainforest trees. "What's your authority, Wally?" Tony asked. "Wally said, 'Tony, you've got to get used to the fact that this is no longer a garden; it's a research tool,' never answering what authority he had," Burgess recalled. "Then I learned, whatever he said had to be done, period."

Not everyone at Biosphere 2 wanted to treat the glass world as a test tube, particularly the few scientists left over from the older days. Steve "Bear" Pitts, a computer modeler, still referred to the Biosphere as somewhere between a machine and a live creature: "She's an exquisite ship, and I really love flying her. I really do." Pitts and colleague Jack Corliss, both hired toward the end of John Allen's reign, spent two years playing with data from the biospherian missions to see what they could learn from modeling the system's evolution as a whole. But that observational approach was not popular with the Biosphere's new managers. Pitts said,

> Wally Broecker had the attitude that there was no valid science in it—he's a reductionist, he comes from the reductionist camp of science, and as a result, he wanted to limit it down to two variables. And you can't do that. Some of the more fuzzy colleagues like Corliss and myself, who actually had some science background, understood that we could learn from the associations that we saw, in this very complex environment, even if we couldn't *prove* anything.

However, Bear Pitts's opinion was not popular at the new Biosphere 2, where quick research results were a high priority. Bear soon earned the distinction of being, in his own words, "the last person Johnny Allen hired, and the first person Wally Broecker fired."

The same attitude change came to Biosphere 2's ocean: Biosphere 2 became a tool to be manipulated. When he arrived to work in the Biosphere 2 ocean in 1995, University of Hawai'i scientist Marlin Atkinson complained, on-site researchers still saw Biosphere 2 as "the Beast" and seemed to believe "Nature is perfect, nature is mystical, if you perturb anything in nature, you ruin it." "I got into the tank, started to turn over rocks to see what was under them, and memos got sent to the director saying I didn't know what I was doing and was going to ruin the coral reef."

Even the vocabulary of the project had shifted. To biospherian Gaie, the coral reef had been "my baby"; to Marlin Atkinson, the former "ocean" was now "the tank." Official Columbia publications referred to the Biosphere as "B2L," shorthand for "Biosphere 2 Laboratory." One 1999 glossy publicity brochure declared,

> The *Biosphere 2 Laboratory*, a magnificent glass-and-steel structure the size of three football fields, is perhaps best known as the mini-planet where four men and four women lived for two years (1991–93) as they sought to determine the challenges of living in an enclosed environment.

In this Columbian reading, Biosphere 2 itself had gone through a fundamental identity change: once a "mini-planet," now it became a "structure," a "laboratory." Biosphere 2 had always been a mix of nature and technology—a cyborg—but now the balance was shifting. Once imagined as a wild, evolving Gaia-like organism, it was now reincarnated—or rather, deincarnated—to become a machine.

The whole concept of Biosphere 2 as a separate world dissolved. No longer would Biosphere 2 breathe its own air; instead it exchanged its liquids and gases daily with the outside world. When Biosphere 2 had been sealed, the carbon dioxide levels in its tiny atmosphere spiked up and down every day, recording the rhythms of photosynthesis as the plants took in CO_2 during the day and stopped consuming it at night. But scientists now protested that such oscillations might negatively impact their experiments. To keep the levels steady, technicians installed a tank to inject CO_2 into Biosphere 2 during the day while plants were consuming it. They installed fans in the South Lung and the rainforest to draw in outside air at night. Scientists also gave up on the old challenge of recycling the Biosphere's water. Now water would pump into Biosphere 2 from wells outside, flow through the rainforest and savanna streams only once, leave the Biosphere for use at the energy center, and finally spray onto the lawn outside for irrigation.

And yet Biosphere 2 continued to model the world outside. For the changing culture of the campus was reenacting, in miniature, the seventeenth-century Scientific Revolution: the shift from experience to experiment, from a belief that humans were part of living nature to a conviction that men should manipulate nature-as-machine. In the new experimentalist vision of Biosphere 2, humans and wild nature were to be separate from each other; humans would observe nature from a distance. The Biosphere's Human Habitat embodied that change. Formerly the biospherian living quarters, now the "Habitat" became a cluster of day-use offices for scientists and technicians. The Habitat was physically cut off from the wilderness by locked doors.

Even the organizational structure running Biosphere 2 reflected a perceived separation between humans and nature. Power over Biosphere 2's ecosystems sat in the hands of people living far away from the Biosphere itself. Local technicians and researchers changed the ocean chemistry and collected data, but often under orders from lead scientists two thousand miles away in New York City. Most of the lead investigators lived near their home academic institutions, meeting weekly only for phone conferences, with notable exceptions such as rainforest researcher Guanghui Lin, who stayed at Biosphere 2 from 1995 onward. Professional scientists, unlike biospherians, wanted to get information out of Biosphere 2, but did not want to be tied to a particular place. As Marlin Atkinson put it, "I want to work in the glass house, but I don't want to be told I have to support the overhead of the glass house with my grants."

The people living close to Biosphere 2 still loved the beast. Younger scientists, working daily at managing Biosphere 2, did describe a close, even devoted relationship to their wilderness, similar to the way the biospherians had once felt. "When the rainforest is stressed, I'm stressed," said rainforest researcher Adrian Southern. But personal feelings aside, he said, his job required that professionally "now when we talk about ecosystem health it's a laboratory, and under this regime we're supposed to manipulate it in any way we see fit . . . It's a living system and it's here to be manipulated." Even some old-time Biosphere 2 construction workers still felt love and devotion to the place. "The stuff that maintenance guys do in there we take very great pride in, because this is where we're at, this is the job we do," said Rod Carender, who had once helped construct Biosphere 2 and stayed on the maintenance staff under Columbia. However, Rod complained, "the research guys that are here now, they don't care about the nuts and bolts of what it takes to get that, to install it, hook it up, to run the cabling, to be safe about it . . . As soon as they get their information, they walk away."

For so long the various managers of Biosphere 2 had wrestled with a fundamental question: how best to control the miniature world? Now the concept of control became key to the Biosphere's new identity as a place for "science." New Biosphere 2 president Bill Harris described the rebirth of Biosphere 2: "We have largely finished the re-engineering of the B2 apparatus and it is now 'science ready.' That is, the facility is controllable and reproducible experiments are possible."[4] In a report describing its research capabilities, Harris again called Biosphere 2 "the world's largest controlled environmental facility." The report emphasized and detailed the possibilities for control: "The unique closure allows whole system manipulations and monitoring that would be extremely difficult in a natural environment. Physical and chemical parameters such as mixing, gas exchange, nutrient concentration, and partial pressure of CO_2 can be independently manipulated."[5]

The Biosphere's seeming controllability lured in collaborating scientists. Marlin Atkinson, experimenting on the Biosphere 2 coral reef, loved that he could tweak the ocean however he wanted in order to study it, he said: "What kept me here was still the facility, that I can enclose a system, I can control the inputs, I can control the outputs." Wally Broecker agreed. To him, the biospherians' human experiment "was a stunt. It had no scientific value," he said. "There's no scientific product because there's no control." To Broecker, tight control of conditions inside the Biosphere would separate stunt from science. When Columbia commissioned a group of prominent scientific consultants, the Washington Advisory Group, to evaluate Biosphere 2 in 1998, the committee similarly concluded in its report that the facility's biggest virtue was its controllability: "Only in a large-scale closed-system facility can significant aspects of ecosystems and their dynamics be studied under controlled conditions." The new stewards of Biosphere 2 seemed transfixed by the same desires as John Allen: they loved the idea of manipulating a whole little world of their own.

There was just one problem with this enticing vision of a controllable little world, however: Biosphere 2, like the world outside, was still a little wild, and not completely controllable. Each scientist wanted to control his or her own piece of the bubble, but everything was too interconnected to control perfectly—particularly when so many scientists had so many different pet experiments. As ocean researcher Heidi Barnett complained, she and her fellow marine scientists felt frustrated because they could not control the ocean chemistry when the rainforest team was manipulating CO_2 levels in the atmosphere overhead: "We kind of had to go along with the CO_2 regimes, we kind of had to deal with the fact that the desert didn't do much, and that the rainforest drove everything . . . Guanghui had priority because he was all gung-ho and had all these people involved in the rainforest, so we're going to go with the rainforest; our research, we're not going to be able to plan what we want to do."

That frustration led the scientists to divide and conquer. In their quest for tighter control, they installed locked doors to separate the Agro-Forestry Biome from the rest of the Biosphere and hung plastic curtains to separate the rainforest from other land areas. In the same spirit of separation, staff members pruned the savanna and marsh to make sure they would not drop leaves into the sensitive ocean. In each area, researchers played with carbon dioxide levels to their own satisfaction and tracked the effects—on cottonwood trees, on rainforest plants, and on corals. In contrast to the biospherians, now the researchers in different areas scarcely needed to speak to each other. Tony Burgess, while working as a professor for Columbia, watched as a new ideology reshaped the little world he had helped design: "We've got to control the levels of CO_2, we've got to control the temperature, we've got to control everything possible, and only vary one thing at a time . . . Well, as soon as you're into that, you don't want this Gaian vision we had of all these complicated biomes combining in some way to get a relative homeostasis of the atmosphere. You want each one chopped off into a separate unit subjected to controlled perturbations."

Instead of a microcosm of planet Earth's wildernesses, Biosphere 2 became a metaphoric representation of science's approach to nature. Its grand pyramids, arches, and tower continued to proclaim the founders' original dreams of monumentality and unity, but its inner divisions of airlocks and plastic curtains mapped out divisions between scientific disciplines.

The new Biosphere 2 was also silently an economic map—a living model of the state of scientific funding. To attract research grants, Columbia's scientists would need to prove themselves quickly. That motivation propelled their program of short-term carbon dioxide experiments. Such controlled experiments were accepted as the quickest route to attract outside funding, said Tony Burgess. "If we don't get good peer reviews for the research, we're finished as a scientific entity. And for that, people are more comfortable with growth chambers. Ultimately they want things they can understand, that are simple, that can be controlled and your variables tightly specified." In Columbia's first years, scientists and administrators talked about knowledge as if it were a crop to be harvested in maximum yields, said Heidi Barnett. "They said it was like oatmeal, the ocean and the rainforest, if we can get something quick and dirty out of the ocean and the rainforest, we'll plant the IAB with wheat, for instance," because "there will be a lag in that" before papers would start to come out.

The need to appeal to outside funders was obvious. Operating the Biosphere 2 campus cost around $10 million per year, nearly $1 million of it in energy costs alone. (Ironically, a facility aimed at studying the Earth's "greenhouse effect" had to burn massive amounts of fossil fuel to deal with its own greenhouse effect; the giant glass house in the desert required a massive cooling system to keep its ecosystems from baking in the sun.) During the first five years of Columbia's management, from 1996 to 2001, Ed Bass had agreed to provide $50 million in support. In the first five years of a new ten-year contract, beginning in 2001, Columbia agreed to put $20 million into Biosphere 2, and Bass agreed to provide $30 million. Administrators hoped that the campus would eventually become self-sustaining through a combination of tourist ticket sales, visiting university students' tuition, and research grants. However, research grants were particularly slow to materialize. A $1 million award from the Packard Foundation in 1999 was the largest contribution.

In its scientific and financial stresses, Biosphere 2 was still modeling approaches toward nature in the outside world. It was not just Biosphere 2 that was struggling to do "quick and dirty" science. As Gaia Hypothesis spokeswoman Lynn Margulis has pointed out, experiments in large-scale ecological complexity are by their nature nearly unfundable. Grant-making categories under the National Science Foundation (NSF) are divided up into discrete disciplines, making it impossible to study interdisciplinary topics and interconnections, such as the links between the atmosphere and life forms:

> The atmosphere, in the organization of NSF, has nothing to do with population biology and physiological ecology . . . If the Earth's lower atmosphere is deeply involved with microbiology and population biology but the charge to atmosphericists is to study the physics, chemistry, and dynamics of the Earth's lower atmosphere, and to ignore all of population biology, how can one submit proposals to study Gaian phenomena and be funded? It is not possible . . . Gaian science is ignored because there is no way to apply for financial support that will involve integrated study.[6]

The bureaucratic and economic structure of academia did not match up with scientists' intellectual understanding of the Earth's interconnectedness. In a 1999 study, ecologist Robert M. May observed that "many of the most intellectually challenging and practically important problems of contemporary ecological science are on much longer timescales and much larger spatial scales" than the scales of most scientific studies. He cited surveys showing that only 25 percent of ecological field experiments addressed a scale greater than ten-meter plots, and only 7 percent of experiments lasted longer than five years, while 40 percent lasted less than a year.[7] The problem was rooted not so much in scientists' minds as in their short research grant cycles. Gaia Hypothesizer James Lovelock, exhorting fellow scientists to join him in studying holistic planetary ecology, hinted at the fate of academic mavericks: "You have nothing to lose but your grants."[8]

Slowly, by the late 1990s, the research climate was changing. In 1999, NSF launched a major Biocomplexity in the Environment initiative to fund research "bringing together interdisciplinary teams of scientists to model the complexity that arises from the interaction of biological, physical, and social systems."[9] In its first few years, the initiative funded the study of all sorts of topics—from land use changes in the Amazon, to chemical cycles in shallow coastline environments, to material flows in urban industrial systems. This new program was one step toward officially acknowledging that the old academic disciplines were too narrow and inadequate to address the complexities of socioenvironmental realities.

Even though Biosphere 2 Center was supposed to be a calm, legitimate research institution now, there was still a quiet air of turmoil about the place. Constant hirings, firings, and reorganizations rattled the staff; the project's own financial insecurity was reflected in the people working there. "There's just too much upheaval," complained ocean researcher Heidi Barnett. "You never know when somebody's going to be fired." After only five years on the Biosphere staff, Heidi said, she felt like an old-timer. In the management of the ocean, agreed Marlin Atkinson, "there have been about five changes of management over four years . . . and the rumors that go with those changes instill insecurity in the people who work here." The presence of well-reputed scientists did not mean an end to ego battles at Biosphere 2. President Bill Harris resigned after a power struggle with a senior scientist in the year 2000. Long before that, Harris had become known for forcing out anyone who disagreed with him. "He likes to micromanage, and he's got a huge ego. He likes to control everything," asserted Wally Broecker, who temporarily quit the project for a year and a half after locking horns with Harris. Of course, many Biosphere staff had similar comments about Wally himself. Wally could seem "very power hungry and he's not very polite," said Bear Pitts, the computer modeler whom Wally had fired; "he says things that could get him in a lot of trouble, if it weren't for the fact that he is the most highly funded individual in the history of the National Science Foundation." Even Michael Crow, Columbia's vice provost, agreed. "Wally is a brilliant quixotic personality, brilliant scientist but lots of ups and downs," Crow said, but rationalized it: "That's pretty typical of great, brilliant scientists."

By 2001, Biosphere 2 was already on its third president in five years. The new president, prominent Australian plant physiologist Barry Osmond, used the metaphor of a playground, as though scientists were children who had to learn to play nicely with each other: "With just one Biosphere 2 available on the planet, it's a huge sandpit for creative people, and Columbia has to, I think, take responsibility for ensuring as many people as possible can play in the sandpit."

When I met Barry Osmond in 2001, he was passionately fighting to make Biosphere 2 succeed. Sitting in his spacious office in shirt and tie, he spoke eloquently about the scientific value of Biosphere 2, which he saw as a brilliant, unique tool for making precise ecosystem measurements—but he knew he was in a race against time to make the project work. "We have about five more years, maybe four now, in which we can actually convince the community," he said nervously, referring to his fellow scientists across the nation. "The majority of the field is still locked into grant programs of $100,000 that they write . . . It's very competitive, almost no funding available, and so it's a think-small environment, a think-small context for think-large questions, and the questions don't get the funding that they deserve."

Barry was an esteemed biologist used to probing small-scale plant processes. Because Biosphere 2 was somewhere between the size of a test tube and a planet, he believed the place could bring together microscopically focused scientists like himself, and connect them to global-change scientists studying the whole world. He fervently believed that Biosphere 2's controllability made it special. He told me, "Everything in the natural environment is hitched to everything else, so this holistic knot is almost impossible to unravel, and this is why facilities like Biosphere are so important—to be able to constrain one thing at a time, under well-measured conditions, where you establish a baseline and then perturb the system. Frankly, scientific inquiry cannot go ahead by any other mechanism."

Barry was excited about possible experiments. At Biosphere 2, he dreamed, scientists could compare the photosynthesis processes in one leaf with the processes observed in the whole rainforest canopy. They could use remote sensing technologies and new algorithms to "make the jump from the leaf to the landscape," he said, using his favorite new catchphrase.

But as this well-spoken plant biologist staked his own scientific credibility on making Biosphere 2 work, he hit against an old nagging question that had never been resolved at Biosphere 2 (or in Biosphere 1): how could the human mind comprehend the true complexity of nature?

On a pleasantly cool desert evening in December 2001, I watched Barry take the stand to rally scientific support for Biosphere 2. At an expense of tens of thousands of dollars, sixty-four scientists had flown to Arizona from national laboratories and universities across the country and the globe. Once again, the high priests of science had been summoned to bless Biosphere 2. Barry was desperately hoping a new research partnership would come out of it. The occasion was a workshop on the future of experimental climate change science, cosponsored by the U.S. Department of Energy (DOE). Biosphere 2 administrators had for years been trying to lure DOE to become a collaborator. Columbia's leaders had tried coaxing government scientists to become involved, and had even prodded members of Congress to lobby the secretary of energy on the subject. This climate change workshop at Biosphere 2, based on a memorandum of understanding signed in the last minutes of the Clinton presidency, was the outcome of that effort.

The conference was in some ways a repeat performance of earlier decades' gatherings in the same spot. In 1984, scientists had flown in from across the country to discuss wild plans of a new world in space. In 1994, researchers had again flown in to propose new experimental directions. But at this gathering in 2001, the focus was much narrower. Only a specific segment of the research community, united by their studies of ecosystem responses to climate change, had been invited. The lecture hall was full of men craning to see each others' graphs and hypotheses projected on the screen. I found myself one of only five women out of the sixty-four invitees; Biosphere 2 and the Department of Energy had kindly flown me in (and rented me a nice gas-guzzling Buick town car) because Barry Osmond believed it could be an historic occasion for climate change science, and he wanted a reporter to document it. My fellow attendees were men attuned to atoms and molecules, not to the quirks of species and living creatures. As one presenter, Paul Falkowski, joked about the abundance of biogeochemists in the room, "They never worry about an organism because they don't know what one looks like."

On the first night, Barry stood tall before his fellow scientists, jovially grinning and declaring in his Australian accent statements similar to those made in the same halls a decade and a half earlier. "I hope you're going to remember it as the place and the time when, hey, that's when Larry said X or when Jerry said Y, and in twenty years' time, I hope you remember this meeting as something that has done something for your discipline and for your students and for our understanding of planet change science," he announced boldly. "The time for large-scale experimental systems such as Biosphere 2 has come." He told his fellow scientists that at Biosphere 2, they could "help define a new discipline" of experimental climate change science. "I'm clearly a little bit starry-eyed here," he admitted. "I'm clearly presuming a great deal. But I think Biosphere 2 is an apparatus in the right hands, at this place, in this time, that will help us provide a new basis for the practice of science."

But his enthusiasm didn't seem exactly matched by his audience. Over the next few days, each invited scientist, all from top government, private, and international laboratories, got five minutes at the projector. Each one proposed a few experiments or research questions for Biosphere 2. But in the in-between discussion periods, the enthusiasm of the old days seemed absent. Many of the visiting scientists were warier about jumping on the Biosphere 2 bandwagon this time around.

I still found it fascinating to listen to these men debate Biosphere 2's merits and flaws, however, because in the process, these cynical visiting scientists were doing more than judging Biosphere 2. They were wrestling with a paradox at the very heart of experimental science: how could one really understand nature through a contrived experiment? This underlying tension was evident in the fact that the scientists could not even agree on their criticisms. From one extreme, critical attendees whined that Biosphere 2 was not controllable *enough* for good science; meanwhile, an opposing camp of critics complained the opposite, moaning that Biosphere 2 was not *natural* enough.

To the first group of critics, the words *control* and *replication* were synonymous with "good science"—to be truly valid, an experiment must be repeated perfectly, under tightly controlled conditions, changing only one variable at a time. But in Biosphere 2, light conditions were constantly changing, and plants were always growing and changing, some scientists pointed out. Other critics complained that perfect replication seemed impossible, because there was only one Biosphere 2 on which to experiment. One respected German researcher rehashed the old proposal to rip out the desert and savanna to create more rainforest, in order to experiment on multiple rainforest parcels at the same time—for example, growing two identical rainforests under contrasting CO_2 levels.

At the same time, another camp took up the opposite complaint: that Biosphere 2 was not *natural* enough for good science. Participant George Hendrey of Brookhaven National Laboratory summarized that concern: "A superb model of plants in a bottle may not be highly relevant to plants in nature." Others chimed in with complaints about Biosphere 2's myriad differences from Biosphere 1: its low light levels, its disproportionately small atmosphere, its "unnatural" soils. "You've got a rainforest growing on an animal feedlot!" one participant exclaimed after learning of the "Wilson's pond soil" that the designers had imported from Arizona cattle ponds. The room convulsed into debate over the revelation that the Biosphere's soils were actually full of cow manure. "It is not something that represents anything in nature; it does not mimic anything," a panel on the soils concluded.

Of course, Biosphere 2's frustrated managers already knew all too well that their little world was not "natural." It was the same old problem that had dogged Biosphere 2's wilderness designers from the beginning: how could the human mind approximate anything as intricate as wild nature? Now the Columbia ocean team constantly worried that their ocean was too different from the ocean in Biosphere 1. The concrete cliffs shaded the seawater for much of the day, which bothered the researchers so much that they considered setting up artificial lights to mimic the tropical sun. (They decided against it because of the high cost.) Indeed, everything about the Biosphere 2 ocean was a little bit off from wild nature, said Heidi Barnett. "You're still not in the tropics, you're not in the equator, you don't have warm flowing seawater . . . It's not just the light and the temperature and all that, it's water flow, it's water chemistry, it's inflow and outflow of species and propagation of things, it's diversity." Thus the new Biosphere scientists and their peers at this latest conference actually were beholden to the same mystique of wilderness that had captivated the Biosphere's designers: somehow they still clung to the hope that their little world must be as natural as possible—even as they simultaneously wanted it to be a machine they could control.

In the Biosphere's land areas, Columbia's terrestrial ecologists fretted over the same issues, said Tony Burgess:

> Is the rainforest representative of any real rainforest? Well, as a botanist I know it's not. There's no real rainforest like it. The insect community is haywire . . . and yet if we openly admit that these things are pretty far off of the continental-scale biome prototypes in terms of their composition, can we extrapolate the dynamics? If we can't do that, we've invalidated the use of Biosphere 2 as a research tool to predict what's going to happen on the Earth.

Of course, the differences were everywhere, because despite the name, Biosphere 2 had not been intended to mimic Biosphere 1 perfectly in the first place. But Columbia's scientists wanted a tool to help them make definitive pronouncements about the outside world. As Wally Broecker put it, "We don't want to only do research that only is applicable to how Biosphere 2 operates, so we've got to do research that is useful to understanding how planet Earth works and predicting how the future will look."

One scientist at the DOE conference summarized much of the confusion over what to do with Biosphere 2: "The problem with ecosystems is that they are large complex ecosystems that we do not understand. So what do we do? We build large complex models—which we do not understand!" He and his peers were hitting on a difficulty fundamental to all experimental science: the tension between embracing too much complexity, and oversimplification. The more one controls an experiment to focus on just the essential facts, the further away one gets from the messy truth of how the real world operates. But the closer one gets to modeling the complexity of that messy real world, the harder it is to draw out any concrete scientific lessons. The Biosphere 2 scientists, and even ecologists as a whole, are not the only ones to have faced this problem; for example, the issue of where to draw the line between oversimplification and excessive complexity has come up in human medical research as well.[10] It is one of the central difficulties inherent within the scientific determination to precisely know the world.

While the scientists debated, another conversation, never officially stated, was still hovering in the shadows of this climate change conference: whatever happened at Biosphere 2, how would this place actually make an impact on the wider world of Biosphere 1? Columbia leaders obviously wanted to influence the outside world—at least, they repeatedly said they did. But even with the prestige of "science" behind them, it would not be a straightforward process.

When I was a university undergraduate, I was among the generation of young environmentalists who found ourselves drawn to this strange glass spaceship in the desert, hoping that someone there would tell us how to save the world. Under Columbia's reign, students arrived from all over the country. We spent months living in apartments and trailers in the little valley below Biosphere 2, pursuing semester and summer courses in environmental studies. The Earth Semester, the most popular program, attracted up to eighty university students each spring and fall, from Columbia and dozens of other schools. We took camping trips into the desert and studied earth science, biology, and environmental economics and policy. Course modules focused on a variety of ecological management issues: Southwestern water resources, species conservation, climate change. Perhaps we were studying the old drama of Eden, recast in scientific-technocratic language: humanity had sinned in its mismanagement of the Earth, and now a new generation must learn to redeem our lost place in the garden. The official guiding principles of the Earth Semester, advertised on the Biosphere 2 website, summed up this outlook:

> Humans are "managing" some significant parts of the Earth system, and for the most part have been doing this unconsciously. We did not set to warm the planet or drive creatures

to extinction; these are unintended consequences of our other activities . . . What do we do about it? . . .

Biosphere 2 offers us some relevant lessons in this regard. Principally:

- The Earth is a materially closed, complex, and dynamic adaptive system.
- Planetary management requires input from many disciplines and the ability to synthesize this knowledge.[11]

Columbia's educators set out to prepare us, then, as aspiring young "planet managers" to redirect Biosphere 1. A year-long Masters of Public Administration (MPA) program at Biosphere 2, run in 2002–3, aimed for this goal as well. Like the Earth Semester, the MPA curriculum cast a sweeping gaze across the planet. A brochure for the program began with a familiar litany of scary environmental catastrophes to come:

The world population is now more than 6 billion. Each day we add another quarter million people to the planet.

In the next twenty years we will need to invest $150 billion to store, clean, and transmit clean water in the United States.

By the year 2100 scientists estimate the temperature of the planet will rise 3.6 degrees F on the average, and sea level will rise by 20 inches.

In 1995 the United States generated 279 million tons of hazardous waste, and in 1998 Americans threw out an additional 298 million tons of household garbage.

The brochure concluded with Columbia's solution: "preparing tomorrow's environmental leaders today / with cutting edge management and analytical tools." It was an age-old vision of apocalypse and salvation, rewritten in new terms: a dark environmental future, to be averted by brave, knowing environmental leaders.

Columbia's vision of the future reflected the new language of environmentalism at the turn of the millennium. Protecting "the future of the planet" had gone from a protest movement to a business, complete with its own professional preparatory programs. On the surface, such developments represented major progress for the environmental movement. But I also noticed that the new Biosphere 2 was less imaginative, less about creating a beautiful "new world" than about stabilizing the old world outside. And as the new Biosphere 2 soon began to illustrate, the meaning of "sustainability" was changing in the hands of major institutions—universities, corporations, and politicians seeking to maintain their own power.

As Biosphere 2's administrators hunted for funding to sustain their own projects, they found themselves lying in bed with some interesting new friends. In 1999, Columbia announced a partnership with the Volvo car company. Volvo would contribute $1 million to Biosphere 2 for one year, and $500,000 in each of five subsequent years. These funds would pay for a new tourist exhibit in the Human Habitat, scholarships for ten students to attend each Earth Semester as "Volvo Scholars," and collaboration between Columbia faculty and employees of the Swedish car company to teach a course unit on industrial ecology to Earth Semester students. In a press conference hooked up by carefully timed video links between Sweden, New York City, and Arizona, Columbia President George Rupp declared, "Both of us are very high quality institutions with a deep concern for the environment." He praised Volvo for its "long history of building high quality cars that are both beautiful and durable and that in particular take special care for their impact on the environment." When asked what Volvo would get out of the deal, CEO Leif Johansson said Volvo was most interested in researching how society would adopt environmentally friendly transportation in the future. "I think if we can do that together with scholars, together with young people who are participating in building a better world in this case, then I think that is a very good benefit for Volvo, much bigger than any type of cash issues that we are discussing," he said. One possible benefit to Volvo that he did not mention, however, became clear later in the day when a class of Arizona sixth graders cut the ribbon on the new exhibit in the Human Habitat: a giant pinball-machine-like metal sculpture with labels discussing environmental problems, funneling into a gleaming bifuel Volvo car at the center. "Having this new exhibit on transportation will be one more very important public education contribution," George Rupp said, referring to the nearly 200,000 tourists who would visit Biosphere 2 every year. However, a giant exhibit about transportation's effects on the environment, featuring a shiny new Volvo car at its center, hinted that corporate-sponsored "public education" was to have its own particular message.

Columbia was not the only university seeking corporate partnerships for environmental research. A few years later, in 2002, Stanford University announced a new ten-year research initiative, the Global Climate and Energy Project, which would devote $225 million to research on energy-efficient technologies. The project would be funded entirely by ExxonMobil, General Electric, and other corporate sponsors—including companies that had also funded a notorious think tank to deny the existence of global warming.[12]

"Sustainability" had at last made it onto major universities' agendas, but it remained to be seen whose version of sustainability would be researched and taught. In the late 1980s, the term *sustainable development* had come into vogue, defined first by the UN Brundtland Commission as "development that meets the needs of the present without compromising the ability of future generations to meet their own needs."[13] But from the beginning other voices had seized the buzzword of *sustainability* for a range of other ecological and economic purposes. Under Columbia, Biosphere 2's administrators took to using the word to refer to their own institutional survival. Chris Bannon described Biosphere 2's financial situation to me: "Over time we want to become self-sufficient. We teach sustainability, we need to practice what we preach." An internal report on the state of Columbia's Biosphere programs declared, "As with any college, research institution or museum, a sustainable financial model must be built around tuition, grants, gifts, and endowment." "Sustainability" in this context just referred to a monetary bottom line.

The push for financial sustainability brought Biosphere 2 into contact with another powerful institution with its own view of sustainable development: the U.S. government. Maverick federal scientists had long flirted with the attraction of Biosphere 2. In the early days, NASA officials and Biosphere 2's creators had visited each others' projects and cosponsored a conference on ecological space technologies. In the mid-1990s, U.S. Department of Agriculture scientists planted wheat in the Biosphere 2 farm. In 2001, the conference between DOE scientists and Biosphere 2 scientists represented the biggest potential for collaboration.

Then, more than ever, Biosphere 2's stewards were hoping for an institutional partner to help pick up the bills.

Department of Energy labs were best known for such inventions as the atomic bomb—so why did the department have a political interest in Biosphere 2? In fact, the DOE's mandate included not only producing energy technologies, but studying the effects of those technologies. For years DOE scientists had researched the ecological effects of atmospheric carbon dioxide and other greenhouse gases, as these gases came from society's energy-generating activities. However, these DOE biologists never enjoyed expensive facilities or congressional clout. Instead they sat and watched as their physicist colleagues, possessing expensive supercolliders and nuclear reactors, raked in more federal grants. Some DOE bioscientists hoped that Biosphere 2 could become a mascot that would help them get a bigger share of the research pot. This hope was partly why, in December 2001, DOE provided $150,000 to fly scientists to Arizona to talk about how the government might benefit from doing research there.

The proposed DOE–Biosphere 2 collaboration came at a key moment for both sides. Columbia scientists and federal scientists both hoped to benefit from political support for climate change science. President George W. Bush would announce his climate change plan a few months later, in February 2002. It would include a new national Climate Change Research Initiative and National Climate Change Technology Initiative, each of which would allocate $40 million for climate change research, split among federal agencies including the Department of Energy.

Bush had just earned anger and ridicule from environmentalists and governments around the world for refusing to sign the Kyoto Protocol, the international climate change agreement that countries had been negotiating for years. He seemed willing to ignore a worldwide scientific consensus that humans' gas emissions were causing global climate change, which could soon bring devastation to ecosystems and people all over the planet. Why then, if Bush seemed determined to keep his head in the sand about the reality of climate change, would he suddenly propose to fund more research on the subject? His climate plan actually enclosed a sick logic. The president was following a well-established pattern: environmental scientists had previously found themselves nicely funded in times when politicians did not want to work to solve environmental problems. In a twisted symbiotic relationship, antienvironmental politicians could use any lingering scientific uncertainty as an excuse to call for "more research"—thereby postponing taking environmental action as long as possible—while scientists themselves would indefinitely propose more research experiments in order to keep the research grants flowing to sustain their own careers. As political scientist Stephen Meyer wrote, "The very characteristics of scientific inquiry that we extol to our students—the careful and methodical (plodding) collection of statistically representative data, the tentative phrasing of conclusions, and the detailed explication of possible sources of error—play easily into the hands of anti-environment forces."[14]

Thus the Bush administration and climate change scientists became strange bedfellows. Even as the U.S. government stubbornly refused to reduce the country's greenhouse gas emissions, it remained one of the largest funders of climate change research. In 2003, U.S. Global Change Research Program funding topped $1.7 billion, split among nine federal agencies. George W. Bush boasted in a speech that his new budget would devote "more than any other nation's commitment in the entire world" to climate change research, insisting, "When we make decisions, we want to make sure we do so on sound science; not what sounds good, but what is real."[15] Bush happily threw around that weighty word *science*. To environmentalists, Bush's words rang with hypocrisy, but the words could have been strangely good news to Biosphere 2, whose financial "sustainability" would depend on society's belief that more scientific experimentation was needed on climate change.

At the conference between DOE and Biosphere 2, as I sat at the dinner table among federally funded scientists, eating delicious catered food paid for by our generous government, the scientists joked about the fact that their research topics had become so detailed and abstruse that they had trouble explaining their work to their own mothers. That somehow seemed a bit worrisome to me, as in theory they were supposed to be using federal funds to prevent major environmental problems. But many of the scientists then began worrying aloud about a related topic. They enjoyed proposing new experiments, but a few confessed to me that they wondered if climate change science might be repeating the history of acid rain research. In the 1980s, two thousand scientists spent half a billion dollars and ten years cataloging the causes and effects of acid rain under the federally funded National Acid Precipitation Assessment Program (NAPAP)—only to watch Congress pass its acid rain legislation before all the science was in. The legislation that passed was weaker than many scientists eventually concluded was necessary; indeed, years later, evidence accumulated that acid rain was still happening.[16] In the 1980s, many scientists studying acid rain had wanted to collect their data without worrying about policy outcomes until the end, preferring to keep research a purely intellectual process.[17] However, as researcher Leslie Roberts argued in *Science*, scientists' predilection to get all the details right, in the service of "good science," played right into the hands of the Reagan administration's desire to avoid dealing with the acid rain problem.[18] The stakes on the acid rain issue were high: as the poisonous rain fell, industrial pollution was killing lake life and damaging forests across the northeastern United States. The stakes on the climate change issue were even higher: rising sea levels could flood coastal settlements and islands, and unpredictable storms, droughts, and heat waves could threaten people and other creatures all over the world, according to most climate scientists' predictions.

Many climate change scientists in the 1990s felt the same way that those working on acid rain had felt, not wanting to hurry their science in order to make quick policy decisions. As Biosphere 2 rainforest researcher Guanghui Lin put it, policymakers seemed to demand of scientists, "Tell me what's going on there and I'll fix it," but as a scientist, he felt that "you can't do that. We need to recognize Earth as a complex system." Still, as Lin observed, "More and more U.S. science and research shows we need to control global CO_2 *now*, not in fifty years." Lin's own research in Biosphere 2's rainforest had already shown that forests could increase their growth to take up excess CO_2, but as global atmospheric CO_2 levels climbed ever higher over the next century, the world's forests would eventually lose their ability to compensate. If the example of Biosphere 2 was to be believed, then the Earth's forests would become saturated as a carbon sink at some point in the twenty-first century. Simply put, trees could not keep soaking up humanity's excess CO_2 forever. The team studying Biosphere 2's ocean, in a study led by Columbia's Chris Langdon, was uncovering similarly dark results: they found that if carbon dioxide emissions continued to grow, assuming a "business as usual" sociopolitical scenario, then worldwide ocean chemistry would be so affected that coral growth would decrease 40 percent by 2065. The dangers that Biosphere 2's corals had faced during the human missions would be repeated in Biosphere 1, Langdon's team predicted. As CO_2 in the air increases, carbonic acid forms in the top layers of the ocean. This acid causes seawater's pH to decline, which in turn decreases the amount of calcium carbonate available to corals to build their skeletons. As Langdon presented his results at the DOE workshop, one alarmed questioner interjected, "If we don't change the direction we're going, forget the coral reefs?!" But then the discussion turned back to the mechanical intricacies of coral accretion and dissolution. Over the years, so much attention at Biosphere 2 had been focused on defining "good science"—yet it was evident that even the best science could not save the world without accompanying action. Scientists already knew the outlines of the predicted ecological apocalypse—yet the call for "more research" continued.

That obsession with accumulating more and more information characterized late twentieth-century science in general. This was the logic of megascience. In one prime example, the huge Human Genome Project received hundreds of millions of dollars in federal government funds, spread over a decade and a half, to map out the entire human DNA sequence. Nobel laureates claimed the genome project would answer the questions "What actually specifies the human organism? What makes us human?" A National Research Council committee argued that mapping the whole genome would "allow rapid progress to occur in the diagnosis and ultimate control of many human diseases."[19] The logic was familiar: maximum knowledge would lead to maximum control and wise management of nature—in this case, the human body. Yet critics complained that this enormous data-gathering project was akin to trying to "list the millions of letters in an encyclopedia without having the power to interpret them."[20]

Ecologists too were gathering massive amounts of data, without certainty as to where it would all lead. Several prominent scientists and conservationists, including Biosphere 2 wilderness designers Ghillean Prance and Peter Warshall, banded together in the year 2000 to initiate an All Species Inventory to catalogue the planet's millions of species before they went extinct. By doing so, scientists hoped to draw public attention to the global extinction crisis. The All Species project's website compared it to the Human Genome Project, insisting that for both data-gathering programs, "the most profound benefits [to the human race] will only be recognized in hindsight."[21]

But would all the new information lead to better action? Harvard biologist Richard Lewontin, an outspoken critic of the Human Genome Project, thought not: "Intellectuals, in their self-flattering wish fulfillment say that knowledge is power, but the truth is that knowledge further empowers only those who have or can acquire the power to use it. My possession of a Ph.D. in nuclear engineering and the complete plans for a nuclear power station will not reduce my electric bill by a penny."[22] Likewise, environmental conservation would not depend on the size of the knowledge pile that the scientists accumulated, but on how they used it.

So Biosphere 2 continued to enclose massive hopes about improving the future of the Earth—and massive uncertainties about what effect the monumental project would have. And in the end, the Columbians, like their biospherian predecessors, had to face hard questions not just about environmental sustainability, but about economic sustainability. When I visited the campus in late 2001, the project's fate was uncertain yet again. Columbia had been in charge of Biosphere 2 for five years and had just signed a contract with Ed Bass to manage it for ten years more. Bass was eager to be free of Biosphere 2, offering to sell the campus to Columbia for a mere $1 million, compared to its $150 million construction cost. But Columbia's administrators remained cautious and instead asked to renew their provisional management contract. The university leaders' decision not to buy Biosphere 2 bespoke some hesitation. Chris Bannon, the Biosphere's second-in-command, was convinced that Columbia was in for the long term. By the end of the new ten-year contract, he pointed out, Columbia would have been at Biosphere 2 for fifteen years, and "knowing Columbia—I work for Columbia now—once they're into something fifteen years, they're a pretty successful place. This place is successful now, but it will be extremely successful by the time that happens," he said. "We've been in the test-bed phase—can we make the Biosphere work, will students come here to this program we've developed? . . . Proof of concept says yes and yes. Now let's start expanding it, taking those steps to move it to the institution that it wants to be by 2010." At the time, in 2001, Biosphere 2 was generating half of its own yearly operating budget through research grants, tourist admissions, and student tuition. Eventually, Bannon and his colleagues believed, that number would increase to 100 percent. "We have a plan in place; we just presented a five-year budget to Columbia that is the most sophisticated and intelligent budget we've ever presented to get us from here to there," he said.

Administrators continued to make long-range plans for the campus's future. In 2001, construction crews completed a brand new multicolored $10 million dormitory complex, with a capacity for two hundred students, right next to the Biosphere itself. Biosphere 2 president Barry Osmond was also talking about big plans. He circulated advertisements for six new tenure-track research faculty positions to be stationed at Biosphere 2, in the fields of plant ecophysiology, biological oceanography, soil science, stable isotope analysis, remote sensing technology, and systems modeling. Looking to the future, he was toying with all kinds of schemes: building alternative energy sources to power Biosphere 2 and making the campus a center for clean energy technologies, or even starting an "academic village" for retired professors who would move to Arizona and become associated with the Biosphere community.

However, other changes were brewing across the country in New York City. George Rupp, the Columbia University president who had first taken on Biosphere 2, left his post in June 2002. In the same month, Vice Provost Michael Crow, who had long championed the Biosphere project, also left, to become president of Arizona State University. Rumors began to spread among Biosphere staff in fall 2002: Columbia's new president, Lee Bollinger, was reconsidering the university's finances, and Biosphere 2 was not on his list of favored projects. The research grant situation was not looking up; the DOE partnership did not seem to be leading anywhere. Even the Earth Semester program, which had attracted growing numbers of college students for several years, was experiencing a decline in enrollment, from eighty students per semester in 1999 to only thirty-two students in spring 2003. (The program's professors blamed the drop on an administrative reshuffling that put New York offices, instead of the Arizona campus staff, in charge of recruiting.) On the whole, after seven years, Columbia's experiment still was falling short of financial stability. Columbia officials abruptly ordered the Biosphere's staff to start running the place with only the funds they generated on campus. But, said Earth Semester professor Tony Burgess, it was clear to everyone, "We weren't bringing in enough to keep the Biosphere alive."

A *New York Times* article made the news official in January 2003: Columbia was abandoning Biosphere 2. Ed Bass sued the university for its "bad faith" in failing to support the educational and research programs adequately, and for backing out of its management contract. Months later, the case was settled, as the university agreed to pay Bass a penalty for the breach of contract. The future remained unclear; once Columbia left at the end of 2003, Biosphere 2 would lie in Ed Bass's reluctant hands again. "In the months to come, we will turn our attention to the future of the property to determine the most viable options for use and operation," said Bass's banker Martin Bowen in a prepared statement.[23] For a castle so beautiful and so expensive, and with so many years of thought and effort invested in it, it was impossible to think of Biosphere 2 sitting unused in the desert, but equally hard to imagine who would take it on next—particularly now that a well-established institution like Columbia had given up. Michael Crow looked on in frustration as his successors abandoned the place. "They were impatient and unwilling to take the necessary risks to advance this important scientific instrument," he complained of Columbia's new administrators. "It wasn't a lack of resources on Columbia's part, it was other people wanting the resources for other things," he said, referring to the university's general turn toward the humanities and social sciences, and away from the earth sciences, under its new president. "We were on the threshold of making this instrument capable of answering some critical questions, just as Columbia pulled the plug," he claimed. But Crow could only watch from outside now.

The miniature world had resisted the aspirations of yet another set of managers. And so it seemed that Biosphere 2's founders had been right about a few things: their creation would model the outside world (even if in unforeseen ways); and it would continue to evolve, beyond what anyone could imagine.

EPILOGUE

The Theater of All Possibilities

IN THE FALL OF 2001 I WAS PLANNING MY SECOND RESEARCH TRIP TO Synergia Ranch. It had been two years since my first visit, and I was eager to hear more stories and get to know people better. But just before I was to leave my home in California to travel to New Mexico, I got a phone call from Gaie. She and some of her companions had sold their "life rights" to an LA movie producer, she explained, and now no one else was allowed to tell their life story. Sorry, but they could not talk to me any more. When I called the Hollywood agent herself, I got the same answer. "It's *my* drama," she snapped into the phone. "You can't steal my drama!" Long conversations with this sharp new owner of "the drama" went nowhere; I would have to cancel my trip.

In the end, after many back-and-forth negotiations over fax and email, everyone finally consented that I could interview people at Synergia Ranch, if I promised not to make a movie or a TV show about them. Oh, and would I mind signing a contract agreeing to change any sections of the manuscript that they found objectionable? Yes, I would mind, I told them. Eventually as a compromise, we agreed that I would consider in good faith their comments on the final manuscript before I published it, which I did. The struggle for control did not surprise me—John Allen and his colleagues had been bashed in the press before, and they could imagine what kinds of stories I would be hearing from others who had been at Biosphere 2. Yet there was also something more going on in the struggle to determine how Biosphere 2 would go down in history. As I road-tripped around the Southwest, interviewing dozens of people who had been involved in Biosphere 2, I realized that the territorial Hollywood producer was not the only one who wanted to own "the drama." Biosphere 2, from its beginning, had always been a grand symbolic construction, a monument to its creators' search for meaning—personal, ecological, and universal. Now they were all still trying to use Biosphere 2 to tell a story that would give meaning to their own lives, and help them make sense of the confused state of the Earth as well.

SANTA FE, NEW MEXICO, 2002

Three years after my first visit, things are feeling decidedly upbeat at Synergia Ranch. As I arrive in early April, the Biosphere's builders merrily show me photos of themselves taken a few days earlier, as they raised their glasses in the sun in a self-mocking celebration of April Fool's Day—commemorating the eighth anniversary of the armed takeover of Biosphere 2. Still creating and running, the group seems to have moved on. At the start of a spring day, the dozen or so current ranch residents and guests gather around the kitchen counters to fry eggs, brew coffee, and eat breakfast. In a quick go-round we each announce our work plans for the day. Everyone seems energetic, cheerful and joking. We head out for our daily tasks. I will do interviews and sift through the archives in the afternoon, but to earn my room and board, I am to join in the morning work on the land as well. Chickens scurry out in the fenced yard behind the buildings. Two women buzz around the orchard in a pickup truck, laying out irrigation pipe to get the young fruit trees through another scorching summer. Sally, once gardener of Biosphere 2, rototills a small outdoor garden plot and hands me seed packets of peas and squash to plant. Laser bops around a big new utility shed, hooking up wires.

Everyone has grown a little older and more professional in the decades since the ranch began. The office has a calm but constant flow of activity about it, as Tango edits the latest Synergetic Press publication on the computer, and Gaie makes phone calls to lab equipment companies seeking donations for the newest project. But some of the old rhythms remain: on Sunday afternoons, ranch residents still enter the geodesic-domed theater for practice time. And on Thursday nights, they still gather at dinnertime in the library, sitting with plates of food on couches and cushions on the floor, to listen to Johnny lecture on metaphysics, monads, dyads, triads, the trajectory of human history, and philosophical systems of the world. The average age in the room is somewhere in the forties now, but all still focus rapt attention on the tall, broad-shouldered seventy-two-year-old man who stands before us, as he makes enthusiastic scribbles with a marker on a paper flip chart. With a grin, a glint in his eye, and an exhorting, triumphant voice, he explains to his audience once again his view of how the world works, and how their work fits into it.

The team has breathed new life into the old network of Ecotechnics projects. The *Heraclitus* is sailing the world's oceans mapping coral reefs, in the name of the Planetary Coral Reef Foundation, a new nonprofit organization headed by Gaie. Mark Nelson, after his experience running the "wastewater gardens" in Biosphere 2, has gotten a PhD in environmental engineering and is spearheading projects to build such gardens in distant locales from Mexico to Indonesia. Every year Ecotechnics associates from all over the world again flock to the Institute's farmhouse estate in southern France for conferences on art, science, and adventure. Three years after the fallout of Biosphere 2, these gatherings resumed in 1997 with a conference on "Complexity: Biospheres, Cultures and Evolution," addressing a range of topics as vast as in the old days: talks on the evolution of the solar system, Mars settlements, vanishing cultures, Eastern martial arts, economic transition in the former Soviet bloc, and others. Again John Allen lectured on grand themes: "The Seven Kingdoms of the Biosphere and the Place of Humanity."

But the team's newest brainchild sits here, small and inconspicuous, on Synergia Ranch. It is a dark red metal cylindrical chamber, just the size of a small bedroom, built so that it sticks halfway out of the dusty red-brown ground. An intricate web of wires, gas monitors, and high-powered lights links the sealed capsule to the high-roofed utility shed next door. This, the former biospherians announce, is their new baby. They call it the Laboratory Biosphere. Built on a tighter budget, the tiny super-sealed chamber is far from the grandeur of Biosphere 2. It is designed to contain no humans, no natural light—just metal walls, powerful grow lamps, a few food crops, and soil. However, the aim behind this little Laboratory Biosphere is more focused, and pointedly ambitious: more than ever, the builders are determined to get off of Planet Earth. Like the Test Module at Biosphere 2, this little prototype is a model for a larger system, a project that the people at Synergia call "Mars on Earth." Unlike the many-faceted Biosphere 2, Mars on Earth will be an exact model for a Mars settlement. The builders tell me they hope to construct it somewhere on Earth first—maybe Australia, they wonder?—then build the real thing on Mars.

Sitting with me on the couch in the colorful Synergia library, Gaie excitedly describes the Mars on Earth vision:

> [We will] put together actually a space-based system to support four people and nothing more. We won't have any corals, we won't have any rainforests, we won't have any savannas or deserts, we'll just have an agriculture-horticulture system with a wastewater garden. And then we'll be asking for participation in robotics, for everything. So eventually, we're hoping, by a decade, it's fine-tuned to where it actually is that system that can be on Mars. And then the next step is to make it out of space materials, and then to attach it to the whole launch program, and look at having it be a part of the 2020 deadline for getting humans to Mars.

Breathless, she bursts out laughing at the loftiness of it all. Actually building a Mars settlement sounds fanciful, from a desert ranch in New Mexico where most of the buildings are handmade adobe. But as they casually discuss the project around the lunch table, the residents of Synergia Ranch point out to me that this is also how they started Biosphere 2: with enthusiastic talks and scribbles on napkins.

Still, despite the optimistic, dream-big tone, something feels less hopeful to me than I imagine it did in the old days. The locus of the new world they imagine now has shifted more firmly than ever away from Earth, and toward the stars. "Are you interested in space, Rebecca?" Laser asks me one afternoon. "No, I'm more interested in Earth," I reply. His rejoinder is quick: "Earth *is* in space." His intuitive perspective on life, in contrast to mine, seems to encompass the whole universe. Yet I wonder whether the team's renewed focus on space represents a real hope for the future, or a desire to escape the messy problems of the home planet.

The next day is a big one for the fledgling Laboratory Biosphere. Laser has hooked up all the electrical wiring; now he is ready to turn on high-powered lamps, which will aim to shine as bright as the sun, strong enough to grow crops indoors, whether on Earth or in space. Everyone on the ranch is busily puttering along in their respective areas, but when the cry comes that the lighting system is ready, all drop their work and come running, cameras in hand. A bunch of us clamber through the door hatch, pile into the ten-meter-square white room, and close the metal door. Laser flips on the lights from outside; as the bulbs pop on one by one, everyone cheers. The powerful tungsten lamps heat up, and we all wriggle down to the metal floor to bask in the warmth. In a playful, theatrical mood, someone pretends to be a little Martian soybean growing under the lamps, and others join in. Gaie stands above the giggling soybeans and muses, "Who shall I eat first?" They squeal in delight: "Me, me, me!" The moment feels upbeat and happy—yet somehow strange to an outsider. I feel surrounded by actors, who take cues from each other in an almost synchronized way, as though they all suddenly decide when is the right moment to be serious, when to be playful, and when to switch back and forth. Still, the synchronous laughter is contagious, and amid the enthusiastic go-go-go spirit, it's hard not to laugh on the floor among fellow soybeans. Somehow the sly spirit of possibility lives on.

Many of the other biospherians and Biosphere-builders, even those long estranged from John Allen, I found still envisioning and trying to create "new worlds." They were charging forward on their own dream projects: two aiming for life in space, one still devoted to caring for the Earth, one still questing to live forever. Taber and Jane, who got married at Biosphere 2 soon after their mission ended, were living together, a decade later, in the house they designed on a quiet street in Tucson. They still worked together as well, at a company they founded, where Taber was CEO and Jane was president: Paragon Space Development Corporation. Together with a handful of colleagues, *they* wanted to be the first ones to grow a plant on Mars, Taber told me. Their mission, their website announced, was to create "an ever-expanding home for life and humanity." In Paragon's headquarters, in a nondescript office park in south Tucson, Jane and Taber greeted me and showed me into the back room. There tiny sea creatures scooted around in aquarium tanks. Sealed glass cages enclosed green shoots of grass and leafy plants under grow lamps. Carefully monitored by electronic sensors, the plants looked something like tiny green hospital patients hooked up to electrodes. Posters decorated the lab walls, displaying photos of the company's early successes: sealed aquatic colonies of plants and invertebrates, which Paragon had already sent into space aboard a space shuttle. These closed systems, Taber proudly pointed out, included the first creatures to complete a full life cycle in space. Next to pictures of these experiments hung visions of the future: artists' renditions of the greenhouse that Taber, Jane, and their colleagues hoped to build on Mars.

Their biospherian friend Linda Leigh remained more grounded on Earth. When I first met Linda at a friend's ranch house north of Tucson, she seemed unsure of her exact destiny, but still determined to live out the biospherian dream of unity between humans and nature. After emerging from Biosphere 2, Linda had completed her PhD on the ecology of the Biosphere 2 rainforest, studying with Howard Odum at the University of Florida. Then she returned to the Arizona desert. I found her living alone in a ranch apartment in Oracle, a few miles up the road from Biosphere 2. At first, she told me, she had dreamed of creating a new center for ecological living in Arizona—the Center for Sustainable Prosperity, she decided to call it. She moved to a large piece of land, far out in the desert on the San Pedro River, to live with some friends. There, they dreamed, they would build their houses with their own hands, grow their own food, generate their own power, and live off what they could grow and sell from the land. But after moving out to the land, Linda found herself always driving back to town to get supplies and see friends. Eventually she decided total self-sufficiency was too much of a stretch, and she moved back to Oracle to work on various environmental and community projects. As we talked about Biosphere 2, she recalled the emotional pain of all the fights in and around the glass world, which had impacted her so intensely that she had given herself ten-year mental deadlines to finally forgive everyone in her mind. Yet she still missed the biospherians' shared passion, she said. "I really miss it. There were parts that were awful that I don't miss, but if I could just take out the parts that were wonderful, I would . . . The group working towards a common goal, with an idea that's crystallized that you're working towards. I miss that so much. And I need to find that again."

Linda, as an ecologist, had long advocated for Biosphere 2 as a model for helping humans understand and manage Biosphere 1. During the biospherians' two-year enclosure, when they were hungry, gasping for oxygen, and squabbling, Linda still tried to find meaning in Biosphere 2 as an ecological metaphor, as she wrote in the public *Biosphere 2 Newsletter*: "Biosphere 2 offers a strong metaphor because consequences of our actions are so obvious. It gives us an opportunity to start to change our thoughts and action through our increasing understanding of the connections."[1] But I was equally struck by a line from one of Linda's journal entries, which she showed me, written around the same time: "I just don't know how people can treat each other so poorly, people with such a strong and beautiful common dream." It was one of the thornier, more painful lessons of Biosphere 2 that years later it still seemed no one could fully explain.

Roy, the wily old biospherian doctor-cum-artist, I found back in Los Angeles, making sense of Biosphere 2 in his own offbeat way. He, more than anyone, seemed content to take his Biosphere experiences at face value and make what art and science he could of them. In Roy's high-ceilinged artist's loft, hidden in a brick building in Venice Beach, a lifetime's worth of adventures and travels lined the walls: paintings of naked women, colorful tapestries, long shelves full of books, double-exposure photographic prints of his fellow biospherians, their bodies melting into photographs of Biosphere 2's natural and technical systems. Roy had already published a half-dozen scientific papers on the effects of the biospherians' calorie-restricted diet, and now he was contentedly working on an artistic video documentary on Biosphere 2. He spent his days sitting in front of screens with two young video artists, poring through hours of video footage that he had shot in the Biosphere. (As the only biospherian still on speaking terms with all seven of his former crewmates, he was also the only one still able to interview them all on camera.) Roy was by far the most acclaimed professional scientist among the former biospherians, having made his career in controlled laboratory science experiments on life extension. Yet, in contrast to his former crewmates, who emphasized to me the scientific merits of Biosphere 2, Roy happily displayed Polaroids of the naked, paint-covered biospherians in his bathroom. On the brick wall outside his home, on a busy boulevard, he proudly hung a giant photographic reproduction of the biospherians' demolished mural, the Butt-Wheeled Wagon of Biosphere 2. He enjoyed the archetype of eccentric trickster and refused to be pigeon-holed; at age seventy-five he posed for an article in the science magazine *Discover* wearing nothing but black bikini bottoms and his signature long white mustache. Roy, like his former comrades, still was crafting his own way of making meaning out of Biosphere 2. His favorite metaphor was, as he called his documentary, "The Voyage of Biosphere 2"—life as an unpredictable journey, full of interesting, unique, and sometimes useful lessons and experiences. And he still did hope that at least history would ascribe meaning to Biosphere 2; at the end of his video documentary, Roy himself faced the camera and intoned in a gravelly voice, "I predict that Biosphere 2 will eventually come to be recognized as one of the most forward-looking and innovative and visionary projects of the second half of the twentieth century."

As I interviewed people from every phase of Biosphere 2's lifespan, from the biospherians to the Columbia leaders, I was surprised to find a distinct commonality: most still found meaning in the Biosphere 2 story by clinging to the concept of this enormous glassed wilderness as a model of the Earth.

At Synergia Ranch, I found that John Allen and his remaining colleagues made sense of their tumultuous experiences by interpreting Biosphere 2's problems as reflections of problems in Biosphere 1. Laser complained of Columbia's management: "They took the people out, they brought in pesticides, and then they took out the soil and put in a monoculture. So there you go, good-bye Biosphere 2. But they're keeping the title. They should really say, 'This is Biosphere 1, we'll just keep on doing what we're doing.'" The creators still imagined themselves as part of a cosmic drama. Biosphere 2, even in its chaos and conflict, still held significance for them as an instructive model of the Earth. Perhaps this was why a few of them were so keen to get to Mars.

At times the Synergia residents seemed to view the Biosphere 2 project as a noble quest sabotaged by outside forces of money and power. In our interviews in 2002, John Allen focused intensely on the way Biosphere 2 had been wrested away from his team. He seemed convinced there was a moral to the story, summarized in the title of one of his poems: "Billionaires Always Win." However, six years later, when he at last published his memoir, *Me and the Biospheres*, his tone had shifted. Perhaps he had forgiven, or at least he wanted to have forgiven. There was not a negative judgment to be found in the book, instead only praise for all of the amazing people he had worked with and known, and celebration of all his life experiences. He continued to try to make meaning of Biosphere 2 by imagining it as a model of the whole world, as he wrote,

> we can see humans first admiring the splendor of Biosphere 2 and the fact that people lived within it; then trying to make scientific reputations and a tourist attraction out of studying its separated wilderness biomes; then putting it up for sale while selling its surrounding area for a scenic housing development. Not that dissimilar from Biosphere 1's history since technological humanity arrived on the scene. Biosphere 2 was a giant human spectroscope.[2]

Furthermore, Allen stretched to turn the analogy between biospheres into a happy, meaningful lesson: "I want present and future humans to know that Biosphere 2 existed, worked and was loved . . . Humans have the power to create new biospheres as well as to restore what has been ruthlessly plundered in Biosphere 1."[3]

This conceptual frame of a "new world" had once motivated people to build Biosphere 2; now it helped them to find peace after all that had happened there. In collecting participants' narratives, I found a common thread within many: to them, Biosphere 2's story was a drama about the fate of the whole Earth and humanity—and therefore, by implication, their lives' work had meaning. Even Michael Crow, the administrator who had led Columbia's takeover of Biosphere 2, took such a perspective. Part of Columbia's original attraction to Biosphere 2 had been the structure's symbolism, he told me:

> We thought that the Biosphere apparatus itself had huge symbolic value as a symbol of limits, if you will, that we all live inside a glass dome, we just don't know it yet. And so why not take the glass dome that we've got at Biosphere and use it as a learning tool . . . Let's use it to answer certain scientific questions, let's take the Biosphere campus and make it into an intellectual center for sustainability-oriented discussions.

After Columbia's reign ended, Crow still sought meaning in Biosphere 2 by seeing it as a symbol for the Earth. He saw the university's abandonment of his pet project as indicative of a wider societal failure to give proper resources to environmental problems, he said. "The Biosphere represents some of the positives and the negatives of where we are culturally, relative to sustainability." He also saw the Biosphere's fate as an indication of problems in the scientific community: "It was sociologically ahead of where the science community was. They weren't ready for big instruments yet." Barry Osmond, Columbia's last president of Biosphere 2, agreed. Barry admitted to me that Biosphere 2 had never won full acceptance from the mainstream scientific community, but he likewise viewed Biosphere 2's woes as lessons about the wider world. He told me that he saw Biosphere 2 as a window on the problematic politics of experimental science, claiming that "a discipline that was seemingly fixed in 'cookie-cutter' mode (with the dough being rolled out in Washington, DC)" prevented scientists from seeing the usefulness of a complex system like Biosphere 2.

Other Biosphere-builders likewise made sense of the drama of Biosphere 2 by interpreting it as a parable about the fate of the Earth—even if the resulting lessons weren't entirely pleasant. Safari, the one-time biospherian candidate and animal worker, believed that Biosphere 2 showed how world environmental problems were rooted in human social problems. She said, "Probably the most interesting subject to come out of the Biosphere is the social situation, because that is what messes up the planet in the first place. Just like Biosphere 2, we study the atmosphere, we study endangered species—and not to say that it's not important, but it's people, policies, and politics that make it all come together." Tony Burgess, the desert designer turned Columbia professor, took a similar perspective. Tony's area of expertise at Biosphere 2 had originally been desert plants; he wanted to study the little ecosystems for detailed natural history lessons. But after twenty years there, Tony seemed more captivated by the project's human lessons. Every year, he lectured Columbia Earth Semester students on "Lessons from Biosphere 2." Though Tony was trained as a botanist, most of the lessons he named in that lecture were social, not scientific: "To survive, your creation, whether biome, research project, or even Earth Semester, must be valued by someone who offers power or wealth," he told the students. Reflecting on the madness of attempts to control both people and ecosystems at Biosphere 2, he offered only humbling advice: "Primate group psychology seems to require and respond to a heroic leader, even if it's an illusion"; yet "one person cannot sustainably manage Biosphere 2, let alone complex systems. We cannot all become heroes, and we should not burn ourselves out trying to be heroic for extended periods."

In our many long conversations, Tony seemed caught between hope and despair—not just for Biosphere 2, but for the Earth itself. He did not flinch away from the harder lessons of Biosphere 2. "The world is run by pathological egos," he flatly intoned to me one afternoon, leaning back in his desk chair and staring matter-of-factly. He even felt Biosphere 2's turmoil in his body. In 1999, Tony learned that he had prostate cancer. He went inside Biosphere 2 with a Sufi healer friend who could read psychic energies; as they stood in the overgrown rainforest, the healer told him, "You're way too hooked into this thing, and it's clear that a lot of your stress that's causing your disease is coming from it." But even as Tony went through cancer treatments, he kept teaching. He still believed that cooperation between humans and ecosystems was possible (even if it was supremely hard), and he felt compelled to pass on what he had learned. For many young environmentalist students arriving at Biosphere 2, Tony was not just a professor, but a woolly-bearded guru in suspenders and a field vest. He seemed determined to convince his students that a healthier planetary future was really still possible, and that we ourselves could create such a future. In his "Lessons from

Desert designer Tony Burgess and architect Phil Hawes together again in the desert, 2007.

PHOTO BY LINDA LEIGH.

Biosphere 2" lecture, he made sure to emphasize some hopeful patterns he had seen in the days of the place's creation—patterns he hoped that students could apply to the Earth around us:

> First: make a personal connection with some part of the Earth. If the connection is to endure long enough for you to gain valued knowledge, you must give it emotion and passion . . .
>
> Understand dynamics of both ecological and social systems, so that you can eventually help them connect and function in tandem . . .
>
> Create a social organization that can sustain your team's purpose and attention. Experience with group dynamics is essential . . .
>
> If you succeed, a dialogue will emerge, and from that interpersonal dialogue, a more encompassing dialogue between you and landscapes, ecosystems and the Earth may develop. This process will inspire great hope.[4]

That lingering idealism of the old days certainly attracted students in Columbia's Biosphere 2 educational programs, myself included. Many students became insatiably curious about the hidden history of this strange-looking "climate change laboratory" in the desert. The more that the Biosphere's administrators tried to downplay the past, the more we glorified it. Aloft in the library tower high above Biosphere 2, we poked among the books to find a few contraband copies of old Theater of All Possibilities plays. One year, some students made T-shirts emblazoned with a diagram from John Allen and Mark Nelson's *Space Biospheres* book, which caused Columbia administrators to go into fits. When I was there, students loved to joke about being part of "the cult" and made souvenir Frisbees emblazoned with the words "this is the disc of all possibilities." In another year, students printed bumper stickers such as "Johnny Dolphin loves you" and "My other car is an adobe spaceship," a reference to the Biosphere-builders' first fantastical model of a floating space biosphere made of Southwestern clay. Perhaps, even as we studied ecological science and politics every day, we loved the mystique of the past because we knew that studying was not enough. We knew that the biospherian missions had gone interpersonally haywire, but we were still drawn to that original spirit lying dormant amid the tangle of plants under huge glass pyramids: the wild hope that even on a troubled and polluted planet, a new world was possible.

The deeper I got into the story of Biosphere 2, the more I found that I, like its many builders and stewards, wanted it to mean something. I wanted it to be a drama with a message, a new ecomyth in a postmodern society sorely in need of ecological inspiration. What would the moral of the story be? That ecological systems are more complex than our rational understanding of them; we cannot completely control even our own creations. That efforts to control nature and to control people are often linked; we cannot get our relationship to the rest of nature "right" without getting human relations "right" as well. That narrow ideas about "good science" continue to dominate the public conversation about our relationship to the natural world. Science, as practiced in current experimentalist "objective" control-based modes, is on its own inadequate as a vehicle for bringing humans into a stewardship role on Planet Earth. But perhaps a middle path is still possible. Perhaps when people feel passionately responsible for their world (and especially when they live intimately in communities as biospherians connected to the land as the basis of survival), they really can remain committed to caring for other creatures. Perhaps we cannot design a perfect world; but perhaps there is still a role for us if we can do as the Biosphere designers hoped: "Listen to the system; see where it wants to go."

As I dug for lessons within the story of Biosphere 2, I realized I was clinging to the same vision as many of the biospherians: the hope (or the illusion?) that this miniworld represented the world outside and therefore was important to history. Of course, that central metaphor animating Biosphere 2 was in some ways a fiction. As author Mary Catherine Bateson has written, "A metaphor can obscure as well as reveal."[5] As much as everyone reiterated the phrase "new world," Biosphere 2 was never its own world at all; it was always tightly plugged into power generators, money flows, and management teams outside. And as much as Biosphere 2's creators wanted to imagine their story as an allegory for the world outside, there was always the dangerous metaphysical possibility that it was not. Savanna designer Peter Warshall put it to me most cynically: "There was nothing avant-garde about what they did. They got a lot of big bucks, they made a totally dysfunctional experimental device that didn't work, and it's all over."

Strange, then, that in spite of it all, I still found inspiration in the biospherians' stories. To the American public, the image of eight starving, out-of-breath, quarreling laborers might be an image of failure—but to me, after meeting those eight, it still seemed an image of another kind of hope. Even on the brink of despair and suffering, these people had remained stubbornly committed to a vision of a self-sustaining ecological system, nearly sacrificing their own bodies to make it work. In them the ancient utopian dream, the dream of a world made new, had lived. Their lives also demonstrated what spurs people to work toward such a world: not just shared theories, but shared stories; not just science, but theater. Columbia's leaders may have officially sought to wash away the Biosphere's past, but ex-biospherian Bernd Zabel also recounted to me how Bill Harris, one of Columbia's presidents of Biosphere 2, once asked Bernd how the old John Allen–led team had inspired such commitment from its employees. Harris was envious because Columbia's leaders never seemed able to generate the same group energy that had propelled Biosphere 2 into being. The Biosphere's original management team illustrated the enduring power of a tribal spirit in modern times. When people feel themselves part of a shared culture that engages their full selves in a balance of thought, feeling, and action, and when they feel that their lives matter to the future of all life, people can work together with incredible energy. Biosphere 2's very existence, in all its problematic glory, was a vivid testament to that truth.

The biospherians reinvented, in their own mixed-up way, what it means to be indigenous to a place and a people, in a postindigenous time for mainstream Western culture. As Martín Prechtel, a North American Indian and Mayan shaman, has written from his experience in crossing cultures, "People have a genuine need to make things with their ingenuity and with their hands . . . Every human being alive today, modern or tribal, primal or overdomesticated, has a soul that is original, natural, and above all, indigenous in one way or another."[6] The biospherians, in their intense, community-oriented approach to life and their care for their created wilderness, actually did try to integrate that spirit of indigeneity with a scientific and global understanding of the world. Perhaps there *was* something really revolutionary about the kind of life they were trying to create together, even if it all turned out a bit nuts. As Prechtel has described, in North America, "Now what is indigenous, natural, subtle, hard to explain, generous, gradual, and village-oriented in each of us is being banished into the ghetto of our hearts, or hidden away . . . our minds are being taught to believe that whatever we can think is actually the center of a person's life . . . a modern culture which thinks with the mind, not with the ancestral soul."[7] The biospherians sought another way.

Indeed, for all the talk of "good science" at Biosphere 2, in every phase of its life the project illustrated that science alone—indeed, rational thinking alone—would not "save the world." Newspaper reporters symbolically destroyed and then rebuilt the project's reputation according to caricatured views of "good science." Yet it was a deep passion for human and ecological connection, not just a scientific outlook, that generated the motivation to actually get Biosphere 2 built. And Columbia scientists, as they accumulated data predicting the frightening future effects of atmospheric change on Earth, found themselves only powerful enough to offer information, not solutions.

By the end of the twentieth century, many environmentalists harbored dark visions of the future. To many observers, the utopian imagination itself seemed dead or dying, as mourned by historians Frank and Fritzie Manuel in their massive survey of Western utopian thought. The Manuels noted that human faith in "all possibilities" seemed to have atrophied in the scientific age:

> What distresses a critical historian today is the discrepancy between the piling up of technological and scientific instrumentalities for making all things possible, and the pitiable poverty of goals. We witness the multiplication of ways to get to space colonies, to manipulate the genetic bank of species man, and simultaneously the weakness of thought, fantasy, wish, utopia. Scientists tell us that they can now outline with a fair degree of accuracy the procedures necessary to establish a space colony in a hollowed comet or an asteroid. But when it comes to describing what people will do there, the men most active in this field merely reconstruct suburbia—garden clubs and all—in a new weightless environment.[8]

Institutions like Columbia University, dreaming of "planetary management," were remaking the utopian vision in their own way, but without the wild spirit of the Theater of All Possibilities. Scientists could explain the world, from big to small—from the unstable global climate down to the molecular reactions causing it—but not at a soul level. Edward O. Wilson, one of the greatest conservation biologists of the late twentieth century, summarized the prevailing scientific view of reality: "All tangible phenomena, from the birth of the stars to the workings of social institutions, are based on material processes that are ultimately reducible, however long and tortuous the sequences, to the laws of physics."[9] Wilson himself, an eloquent spokesman for conservation, took this materialist view. But there was still a gap in science's mode of explanation. As metaphysicist Ken Wilber put it, "Science tells us about electrons, atoms, molecules, galaxies, digital data bits, network systems: it tells us what a thing is, not whether it is good or bad, or what it should be or could be or ought to be."[10]

Biosphere 2 floundered in the world of official science in part because it was built not just on science but on symbolism—on a melding of myths, from the Garden of Eden, to the wild frontier, to John Allen's space-age fantasy plays. The myths animating the Biosphere-builders were stories about restoring a sense of purpose to humanity's role in the natural world. *Myth* is a curious word, often taken to mean a "falsehood" or "lie"—myths, to scientifically-minded people, are untrue, antiquated stories, believed only by primitive societies. Yet many scholars of myth have noted the crucial role that myths have played in nearly every society except our own: they explain and give meaning to the universe and human life. The great mythologist Joseph Campbell has described four functions of myth, common to nearly all cultures: opening a mystical sense of wonder toward the universe; explaining how that universe works; supporting a particular social order; and telling people how to live. But modern rational society, for all its scientific knowledge, has turned away from the mythologies of traditional cultures without offering new myths to replace them, as Campbell has argued: "Today there are no boundaries. The only mythology we have today is the mythology of the planet—and we don't have such a mythology . . . The models have to be appropriate to the time in which you are living, and our time has changed so fast that what was proper fifty years ago is not proper today."[11] The Theater of All Possibilities actors seemed willing to try on every possible myth on stage, to create their own synergetic philosophy appropriate to a time of scientific and global consciousness. The drama that they lived at Biosphere 2 was their own mixed-up creation myth, a contemporary reenactment of the archetypal quest for a new world. Perhaps the biospherians were also living out a new creation/destruction myth for the rest of us—a myth appropriate to a time caught between nostalgia for wilderness and lust for technology, and caught between cosmic hopes and apocalyptic fears for the planet's future.

In other, more personal, ways I unexpectedly found myself sucked into the drama of Biosphere 2. This happened in my friendship with Roy Walford, who at first seemed one of the least scarred by Biosphere 2, but may, in a bitter irony, have suffered the most in the end. Before Biosphere 2, Roy had devoted both his scientific career and his personal life to trying to make life longer. He had researched the effects of calorie-restricted diets in mice (and then in himself and his fellow biospherians) because he believed such diets could extend the natural lifespan. He undertook wild adventures, such as the biospherian mission, to make life seem longer—extending his lifespan psychologically. But that last adventure may have had the opposite effect. In the years after exiting Biosphere 2, Roy began suffering from symptoms of amyotrophic lateral sclerosis (ALS), commonly known as Lou Gehrig's disease, causing him to gradually lose the use of his muscles. Though he could not know for sure, Roy wondered if the disease was linked to his two years spent breathing the air of Biosphere 2, whose high nitrous oxide and low oxygen levels would have caused the death of brain cells. The first time I met Roy in 1999, his voice was quiet and gravelly from his weakened throat muscles. Over the next few years, I would visit him in his colorful studio loft whenever I passed through Los Angeles. Soon Roy was walking with a walker, wearing a neck brace to hold his head up, using ropes of webbing strung across his apartment so he could pull himself around on his feet. Meanwhile he kept abreast of the latest medical research, hoping to make himself a guinea pig for new experimental treatments.

In the spring of 2004, I got an email unlike any other I had ever received. It was from Roy. He was running out of time to live, he wrote, and had made plans to be cryogenically frozen at the moment of death. No one had ever woken up from such a procedure, but doctors had agreed to try out a new freezing technique on him. Roy's body would wait in cold storage for about fifty years, or until a cure for his disease was found. Then he got to the point of his email: when he woke up in fifty years, most people he knew would already be dead, which would be a bit of a shock. I was one of Roy's youngest friends, so, he wanted to know, could I promise to be there when he woke up? I was twenty-five years old at the time and had trouble imagining myself at age seventy-five. But Roy could. While the other biospherians were trying still to transcend life on Earth, he was trying to transcend time—by surpassing the natural lifespan of his own body.

Soon after, on April 27, 2004, Roy died. As it turned out, he was not frozen. During the death process, he lost oxygen flow to the brain and went into a coma, causing brain damage beyond the level he had deemed acceptable in order to be cryogenically preserved. Instead he was cremated and his ashes scattered at sea. He was seventy-nine—not young by most standards, but not the 120 years he had hoped for.

ORACLE, ARIZONA, 2004

When I get back to Arizona for Roy's memorial service, I go first to see my old mentor and friend Tony Burgess. Now that Columbia has finally abandoned Biosphere 2, Tony is out of work. He's getting ready to leave Arizona after twenty years spent working at Biosphere 2 and is moving back home to Texas, packing up his potted cacti and his boxes and boxes of files. We get into his little green hybrid car, with its bumper sticker declaring his creed—"Tree-Hugging Dirt Worshipper"—and zoom up the desert highway one last time to Biosphere 2.

Roy's memorial service turns into a reunion. Old friends stand around the long oval conference table at empty Mission Control, chatting and hugging. Ex-biospherians Taber, Jane, and Linda are there, as are Biosphere 2 architect Phil Hawes, Mission Two biospherians Bernd and Tilak, and a room full of about twenty friends, many of whom poured their lives into building the Biosphere. Everyone is older of course, some a little heavier, some a little grayer. They greet each other warmly, exchanging updates on their lives and latest projects.

With Columbia's staff and students gone, the Biosphere 2 campus feels like a ghost town now. Ed Bass still owns the place, but no one knows what will happen next. A few lingering employees operate a visitors' center, charging each tourist $19.95, the campus's only source of income. The rest of the campus sits eerily silent—labs, offices, lecture halls, Mission Control, and even Columbia's brightly-colored, brand-new $10 million student dormitory, all abandoned. One former employee whispers to me that he has heard that working there now is "like working in a mortuary."

We have received special permission to enter Biosphere 2 one last time, but we have to be out by 5 P.M. Inside, we sit in a circle on the carpet in the high-ceilinged second-floor plaza of the Human Habitat. People tell stories about Roy, and about how he cheered up his comrades inside Biosphere 2. Lying on the floor, Jane reminisces aloud how on this same carpet, the eight biospherians used to come together for solstice feasts: "We spent hours laying around on this very floor, laying on this very floor, stuffing ourselves to the gills," some of the rare occasions when "we were usually decent to each other." She recalls how even during the darkest times, Roy seemed to maintain a balanced attitude inside the Biosphere, even when "it could have been really easy to sink into a complete state of despair." Linda tells how Roy constantly asked his companions for samples of their bodily fluids, whimsically carved papayas with a scalpel, published numerous scientific papers, and made his crewmates eat a terrible cold soup of sweet potato leaves. "On a scale of one to ten with 'one' being unbearable and 'ten' being fabulous, our lives in the Biosphere encompassed the entire range," she says.

Indeed, the occasion feels not only like a memorial for Roy Walford, but a memorial for Biosphere 2 itself. We get up from the carpet and stand leaning at the glass doors to the Habitat balcony, looking down at the once-lush farm fields. Now there is nothing but pale, dry brown dirt. "That's where I spent almost all of every day," sighs Jane, leaning on the glass, shaking her head. After years of such devotion, followed by such anger, and now just a lingering sense of loss, there is nothing to do but shrug and laugh. "Where's the farmer?!" demands Linda, joking; the others start needling Tilak, the Mission Two biospherian in charge of agriculture, to get back down there and get to work.

In a sort of funeral procession, we wander the basement tunnels and wilderness areas together. The creators scrutinize plants and little details along the way. As we stand on the tourist boardwalk in the savanna, Taber leans over the concrete cliffs, squinting down at the ocean below, frowning at a layer of algae scum floating on the surface. Safari and Jane discuss why some mangroves seem to be doing better than others down in the marsh at the ocean's edge. As we walk, everyone is checking and exclaiming over their favorite plants. Tony Burgess admiringly strokes the thick trunk of a Queensland bottle tree growing toward the glass roof in the savanna. "It was this tall when I planted it," he says proudly, holding up his hands. We climb downstairs, duck into a basement corridor, walk down the hall, climb up again, and emerge in the rainforest. Everyone bends to inspect the heavy leaf litter covering the forest floor. Vines cover the trees; the rainforest is overgrown and less diverse than it once was, but still thick and green. "Crazy ants" scurry in thick processions over concrete stairs, along vines and window panes; they, at least, still seem to enjoy paradise in Biosphere 2. Tony thinks the plants may be in heat shock, he says, because the current managers are skimping on the Biosphere's air-conditioning budget; on the other side of the glass, it is early June in southern Arizona, with searing one hundred–degree days. Still, when I ask Tony what he thinks of the created wilderness now, he shrugs, looks around, and nods his cautious acceptance: "It's still alive."

At last we pass through the long, white tubular underground tunnel to Biosphere 2's South Lung. The rubber ceiling liner sags, because the lung is no longer being used to maintain air pressure in the Biosphere, but even with the black neoprene sheet drooping over our heads, the space still feels huge and cavernous. A friend stands alone at the lung's edge, by the curving wall, playing a soaring elegy on the violin. Others walk around quietly, conversing arm in arm. But at last it is closing time. We step out the lung's side door into the sharp desert light, the sun glowing orange over the mountain ranges to the west. As we walk away, looking up at the towering glass pyramids full of plants, former biospherian Jane wonders aloud what many must be thinking: with such a magnificent structure, why can't anyone think of anything profitable to do with it? No one has an answer. "No one has asked Biosphere 2 the right question yet," muses her husband, Taber. Old friends exchange hugs and goodbyes, get in their cars, and drive back out to the highway, leaving Biosphere 2 all alone.

After my many inquiries into the Biosphere's past, the mystery in the equation continued to be Ed Bass. Why had he willingly put up $250 million of his own money to build and run Biosphere 2? And what was he thinking now? Unfortunately, Ed wasn't interested in talking. Unlike the rest of the characters—who seemed glad that someone was writing the story of Biosphere 2, or were at least willing to chat—Bass was protected on all sides by professional handlers and associates. They repeatedly turned down requests for interviews. At one point I managed to sit next to Ed at a conference at Biosphere 2. He seemed an unassuming guy: quiet demeanor, silver hair, cowboy boots, and a Bio2 baseball cap. He sat alone, listened to scientists talk, and gave his own short speech about the original goals of Biosphere 2, encouraging Columbia scientists to "find the direction up the mountain" for the experiment, and in the meantime to enjoy "the beauty of nature itself" on the Biosphere campus. But when the evening's talks were over, I scarcely got a few moments to speak to Ed before a ring of the Biosphere's Columbia leaders swooped in to shake his hands, then ushered him to a waiting golf cart outside, where he was quickly whisked off into the night.

Still, a few years later, I managed to track Ed down at a gathering of environmental bigwigs at his alma mater, Yale; he was chair of the Leadership Council of the Yale School of Forestry and Environmental Studies, where I was a graduate student at the time. Sidling up to Ed at a reception full of men in suits, with the help of another silver-haired eco-millionaire who had befriended me, I introduced myself to him again. But again Ed was reticent, almost hesitant to speak about the project that had consumed such big chunks of his life and fortunes. All he would say about Biosphere 2's future was, "We're in a holding pattern." He seemed relieved when someone else changed the subject. Ed did not seem ill-willed, but he was clearly uncomfortable being the center of the attention, or being in charge of the fate of Biosphere 2.

It soon seemed, however, that Ed Bass just wanted to be free of the project. In March 2005, a group of scientists convened at the National Academy of Sciences in Washington, DC, for yet another meeting about what to do with Biosphere 2, but though various federal agencies and universities expressed interest at the time, nothing came of it. Meanwhile, a new Biosphere 2 website offered the property for sale to the highest bidder, as an "exceptional real estate opportunity." Sketches by an Arizona design firm, hired by Bass, offered a list of suggestions and pictures for prospective developers of the property: they could turn Biosphere 2 Center into a scientific research park, or a senior citizens' colony, or a "Nike town," or recreational facilities—an "extreme hunting facility," a "health spa," or perhaps a "fat farm." The real estate website featured a variety of artists' sketches of a Biosphere 2 surrounded by swimming pools, hotel rooms, palm trees, and cabanas. The lists and sketches almost seemed like a joke, until the news came out in February 2006: Arizona newspapers announced that a major real estate developer, Fairfield Homes, was purchasing the Biosphere 2 property, with plans to turn it into a suburban housing subdivision, just like the dozens of new cookie-cutter neighborhoods sprawling into the desert in every other direction outside of Tucson. The new developers were more interested in the land's real estate value than they were in the castle of Biosphere 2 itself. "We've been looking at the Biosphere and been trying to integrate. There's probably no way of making those building code requirements work, so we're probably going to be shutting that down," said a spokesman for the development company in reference to the Biosphere 2 structure. "If there's anything we can save and utilize, we'll save it."[12]

A few months later, the *New York Times* printed the Biosphere-turned-housing-development story. Interestingly, the *Times* real estate writer too sought meaning in Biosphere 2 as metaphor for Earth: "The Biosphere was designed to simulate the Earth's environment. By succumbing to sprawl, it may have done just that."[13] At that point, many of us who had been involved with Biosphere 2 assumed it was all over. Fortunately, the *New York Times* got it wrong. The real estate deal eventually fell through, and meanwhile, behind the scenes, Ed Bass's representatives were in discussion with University of Arizona (UA) leaders in Tucson. After a year and a half of workshops, the university's earth scientists began managing Biosphere 2 in mid-2007. Another company, created by Ed Bass and his associates, bought the land, and the Philecology Foundation—another of Bass's creations—pledged a $30 million gift spread out over the next ten years to support the university's programs there.

The new program included multiple arms. At a conference center called the B2 Institute, UA set out to host symposia on "Grand Challenges" such as climate change, global energy, and water supplies—and so the hope of Biosphere 2 as a site for working out the planet's future continued. Meanwhile, the UA science team began designing new experiments. In contrast to Columbia's intense focus on carbon dioxide and climate change, the UA team seemed less wedded to a single idea, "waiting for the right question to come along," said Travis Huxman, director of the new B2 Earthscience program. In late 2007 the university received its first external grant for Biosphere 2, $2.5 million from the National Science Foundation to study what might happen to tropical rainforests under global climate change, comparing Biosphere 2's rainforest with existing research sites in the Amazon. Huxman and his colleagues also were designing a new minilandscape, building a hillslope in the biospherians' old farm, to track the flow of water as it moved through the system. Now UA's water study would aim once again at working out an "experimental complexity compromise," Huxman said. "You can have the complexity of the real world, but it's controlled enough so you can make a measurement." And so returned the age-old challenge of Biosphere 2: how to find a compromise between complexity and control. Might this new team of scientists find a way?

The Biosphere's new leaders also talked of engaging the public in science through tours and events. UA student researchers at the new "B2" would have to talk with the public as well. In the process, the students would gain a kind of education not available to most young scientists: they would have to learn to make their work relevant to society, Huxman said. At least, that was the hope. As I recalled meeting Department of Energy climate scientists at Biosphere 2, who had joked together that they could not explain their science to their own mothers, this at least sounded like a promising step.

Contemplating the unknown future of my own biosphere, Biosphere 1, I continue to look for meaning in myths, in the stories people tell me and in the stories that I tell to others. During one of my visits to Synergia Ranch, Laser asked me if I would have liked to have been a biospherian. I laughed. I certainly did not envy the hungry biospherians' survival struggles, or the rest of the chaotic interactions that pulled their project apart. Yet I found that I did envy their original shared belief in a common dream—their real faith in "all possibilities," that together they could create their own script, even their own world. Indeed, I would like to be a biospherian—of Biosphere 1. We're going to need a big crew.

NOTES

SEEDS

1. John Allen, *Me and the Biospheres* (Santa Fe, NM: Synergetic Press, 2008), 30.
2. Phil Hawes, "Transdisciplinarity and the Biosphere 2 Project" (paper written for the First World Congress of Transdisciplinarity, Setúbal, Portugal, November 3–6, 1994).
3. Benjamin Zablocki, *Alienation and Charisma: A Study of Contemporary American Communes* (New York: Free Press, 1980), 49.
4. Ram Dass, *Be Here Now* (New York: Crown Publishing Group, 1971), 104.
5. Yaacov Oved, *Two Hundred Years of American Communes* (New Brunswick, NJ: Transaction Books, 1988), 481.
6. John Allen to *Village Voice*, April 2, 1991.
7. Laurence Veysey, *The Communal Experience: Anarchist and Mystical Communities in Twentieth-Century America* (Chicago: University of Chicago Press, 1978), 291.
8. Ibid., 279.
9. Hawes, "Transdisciplinarity."
10. Allen, *Me and the Biospheres*, 52.
11. Ibid., 13–14.
12. Ibid., 82.
13. Ibid., 27.
14. Johnny Dolphin (John Allen), *Liberated Space* (Santa Fe, NM: Synergetic Press, 2000), 7.
15. Ibid., 9.
16. Stephen Storm, audiotaped interview by unknown interviewer, Oracle, AZ [c. late 1980s]. Lent to author by private owner.
17. Dick Fairfield, ed., *Modern Utopian*, quoted in *Whole Earth Catalog*, Fall 1968, 41.

18. Louis J. Kern, *An Ordered Love: Sex Roles and Sexuality in Victorian Utopias* (Chapel Hill: University of North Carolina Press, 1981), 315.
19. Todd Gitlin, *The Sixties: Years of Hope, Days of Rage* (New York: Bantam Books, 1993), 420.
20. Johnny Dolphin, *Tin Can Man,* in *The Collected Works of Caravan of Dreams Theater,* vol. 2 (London: Synergetic Press, 1984), 121.
21. A. R. Orage, foreword and translator's note, *Meetings with Remarkable Men,* by George Ivanovich Gurdjieff (New York: Penguin Arkana, 1985), vii–x.
22. John G. Bennett, *Is There "Life" on Earth? An Introduction to Gurdjieff* (New York: Stonehill Publishing, 1973), 23.
23. Dolphin, *Caravan of Dreams Theater,* 2:1.
24. John Allen to author, September 10, 2008.
25. Rosabeth Moss Kanter, *Commitment and Community: Communes and Utopias in Sociological Perspective* (Cambridge, MA: Harvard University Press, 1972), 54.
26. Frank E. Manuel and Fritzie P. Manuel, *Utopian Thought in the Western World* (Cambridge, MA: Harvard University Press, 1979), 1.
27. Ibid.
28. Donald E. Pitzer, ed., *America's Communal Utopias* (Chapel Hill: University of North Carolina Press, 1997), 6.
29. Priscilla J. Brewer, "The Shakers of Mother Ann Lee," in Pitzer, *America's Communal Utopias,* 37–38.
30. Pitzer, "The New Moral World of Robert Owen and New Harmony," in Pitzer, *America's Communal Utopias,* 88.
31. Zablocki, *Alienation and Charisma.*
32. *Modern Utopian,* quoted in *Whole Earth Catalog.*
33. Veysey, 466.
34. Jane Poynter, *The Human Experiment: Two Years and Twenty Minutes Inside Biosphere 2* (New York: Thunder's Mouth Press, 2006), 15.
35. Storm, audiotaped interview.

CREATE AND RUN

1. Allen, *Me and the Biospheres,* 56–58.
2. Ed Bass, audiotaped interview by unknown interviewer, Oracle, AZ [c. late 1980s]. Lent to author by private owner.
3. John Allen, *Succeed: A Handbook on Structuring Managerial Thought* (Oracle, AZ: Synergetic Press, 1985), 3.
4. Lewis Mumford, *Technics and Civilization* (New York: Harcourt, Brace, and World, 1934), 353.
5. Mark Nelson to author, December 4, 2006.

6. John Allen, Tango Parrish, and Mark Nelson, "The Institute of Ecotechnics: An Institute Devoted to Developing the Discipline of Relating Technosphere to Biosphere," *Environmentalist* 4 (1984): 205.
7. Poynter, 24.
8. Allen, *Me and the Biosphere* (draft manuscript, 2002), 36. Lent to author by private owner.
9. Storm, audiotaped interview.
10. Storm, audiotaped interview.
11. Dolphin, *Tin Can Man*, 145.
12. Allen, Parrish, and Nelson, 205.
13. William S. Burroughs, "The Four Horsemen of the Apocalypse," in Institute of Ecotechnics, *Man, Earth, and the Challenges* (proceedings of the Planet Earth Conference, Institute of Ecotechnics, Aix-en-Provence, France, December 12–15, 1980) (Santa Fe, NM: Synergetic Press, 1981), 163.
14. U.S. Congress, *The Wilderness Act*, Public Law 88–577, 1964.
15. Dolphin, *Tin Can Man*, 122.
16. Stewart Brand, ed., *Whole Earth Catalog*, 1968.
17. Mark Nelson, "Conceptual Model to Evaluate Forces Operating on Planet Earth," in Institute of Ecotechnics, *Man, Earth, and the Challenges*.
18. Ibid., 269, 277.
19. Allen, *Succeed*.
20. George Ivanovich Gurdjieff, *Beelzebub's Tales to His Grandson: An Objectively Impartial Criticism of the Life of Man* (New York: Viking Arkana, 1992), 1085.

GENESIS

1. Jeanne Marie Laskas, "Weird Science," *Life*, August 1991, 71.
2. John Allen and Mark Nelson, *Space Biospheres* (Oracle, AZ: Synergetic Press, 1989), 4.
3. David Ansley, "The New World," *Discover* 11, no. 9 (September 1990): 60.
4. Gina Maranto, "Earth's First Visitors to Mars," *Discover* 8, no. 5 (May 1987): 28.
5. Maxwell Hunter, quoted in Michael A. G. Michaud, *Reaching for the High Frontier: The American Pro-Space Movement, 1972–84* (New York: Praeger, 1986), 295.
6. Michaud, *Reaching*, 104.
7. Linda T. Krug, *Presidential Perspectives on Space Exploration: Guiding Metaphors from Eisenhower to Bush* (New York: Praeger, 1991), 68, 73.

8. National Commission on Space, *Pioneering the Space Frontier* (New York: Bantam Books, 1986), 2–3.
9. Ric Volante, "Local Scientists Developing New Kind of World Near Oracle," *Arizona Daily Star* [c. 1987].
10. Dominique Parent-Altier, "Biosphere 2," *'Scape*, June/July 1990, 36.
11. Jim Robbins, "Visitors to a Small Planet: Earth, Meet Biosphere II," *New York Times*, October 18, 1987.
12. Quoted in Irving Rappaport, *Will Apples Grow on Mars?: The Glass Ark*, produced by Landseer Film and TV Productions Ltd., 1991.
13. Ibid.
14. Roy Walford, quoted in Thomas H. Maugh II, "Roy Walford, 79; Eccentric UCLA Scientist Touted Food Restriction," *Los Angeles Times*, May 1, 2004, B15.
15. Quoted in Rappaport.
16. Charles Bremner, "Closed Encounters in the Desert," *London Times*, March 4, 1989.
17. Ghillean Prance, quoted in John Allen, *Biosphere 2: The Human Experiment* (New York: Penguin Books, 1991), 40.
18. Space Biospheres Ventures, "Biosphere 2: A Project to Create a Biosphere," press release, Oracle, AZ, March 7, 1991.
19. Allen, *Biosphere 2*, 33.
20. Christopher Thacker, *A History of Gardens* (Berkeley: University of California Press, 1985), 253.
21. "Duke Gardens in New Jersey," http://www.njskylands.com/atdukgar.htm (accessed March 2004).
22. John Prest, *The Garden of Eden: The Botanic Garden and the Re-creation of Paradise* (New Haven, CT: Yale University Press, 1988), 42.
23. Caravan of Dreams Theater, *Faust*, in *The Collected Works of Caravan of Dreams Theater*, vol. 1 (London: Synergetic Press, 1984), 125–26.
24. Tony Burgess, "Lessons from a History of Biosphere 2" (lecture to Biosphere 2 Earth Semester students, Oracle, AZ, October 25, 1999).
25. Linda Leigh, quoted in *Biosphere 2: Noah's Ark for Mars?* video [c. 1991]. Lent to author by private owner.
26. Quoted in Rappaport.
27. Ibid.
28. Lynn White, "The Historical Roots of Our Ecologic Crisis," *Science* 155 (1967): 1203–7.
29. Candace Slater, "Amazonia as Edenic Narrative," in *Uncommon Ground: Rethinking the Human Place in Nature*, ed. William Cronon (New York: W. W. Norton, 1996), 488–90.
30. Evan Eisenberg, *The Ecology of Eden* (New York: Alfred A. Knopf, 1998), xv.

31. Carolyn Merchant, *Reinventing Eden: The Fate of Nature in Western Culture* (New York: Routledge, 2003).
32. Francis Bacon, *The New Organon*, ed. Fulton H. Anderson (New York: Macmillan, 1960), 267.
33. Peter Warshall, "Lessons from Biosphere 2: Ecodesign, Surprises, and the Humility of Gaian Thought," *Whole Earth*, Spring 1996, 27.
34. Linda Leigh, "Biosphere 2 Report from the Inside," *Buzzworm* 3, no. 6 (November/December 1991): 23.
35. Quoted in Rappaport.
36. William Bradford, quoted in Roderick Nash, *Wilderness and the American Mind*, 3rd ed. (New Haven, CT: Yale University Press, 1982), 23–24.
37. John Muir, *Our National Parks* (Boston, 1901), 74; and Muir, "God's First Temples: How Shall We Preserve Our Forests?" *Sacramento Record-Union*, February 5, 1876. Quoted in Nash, 125, 130.
38. Nash, xii.
39. David Brower, foreword to *The Sacred Earth: Writers on Nature and Spirit*, ed. Jason Gardner (Novato, CA: New World Library, 1998), xiv.
40. Donald Worster, "The Ecology of Order and Chaos," in *The Wealth of Nature: Environmental History and the Ecological Imagination* (New York: Oxford University Press, 1993), 167.
41. U.S. Congress, *Wilderness Act*.
42. N. Katherine Hayles, "Simulated Nature and Natural Simulations: Rethinking the Relation between Beholder and the World," in Cronon, *Uncommon Ground*, 410.
43. William Mitsch, preface to *Biosphere 2: Research Past and Present, Ecological Engineering* 13, nos. 1–4 (Elsevier Science, June 1999).
44. Mark Nelson, "I Live in a Glass House," in *Life Under Glass: The Inside Story of Biosphere 2*, by Abigail Alling and Mark Nelson (Oracle, AZ: Biosphere Press, 1993), i.
45. William Dempster to author, September 10, 2008.
46. Poynter, 103.
47. Karen Armstrong, *A Short History of Myth* (New York: Canongate, 2005), 2.

THE POWER OF LIFE

1. Edward Stiles, "Space Survival 'World' Is Born," *Tucson Citizen*, January 31, 1987.
2. "M. Augustine CV," Global Ecotechnics Corporation, http://www.biospheres.com/keyaugustcv.html (accessed November 2005).
3. Allen, *Biosphere 2*, 29.

4. Clair Folsome, "Background for Materially Closed Ecologies" (printed transcript of speech at Biospheres Conference, Oracle, AZ, December 8, 1984). Lent to author by private owner.
5. Allen, *Biosphere 2*, 29.
6. Kevin Kelly, *Out of Control: The Rise of Neo-Biological Civilization* (Reading, MA: Addison-Wesley, 1994), 148.
7. Allen, *Succeed*, 3. Emphasis in original.
8. John Allen, quoted in phone interview with John Brockman, 1989, audio tape. Lent to author by private owner.
9. David Stumpf, quoted in Peter Aleshire, "Scientists Curious but Cautious," *Arizona Republic*, October 1, 1990.
10. John Allen, "Historical Overview of the Biosphere 2 Project," in *Biological Life Support Systems*, ed. Mark Nelson and Gerald Soffen (proceedings of the Workshop on Biological Life Support Technologies: Commercial Opportunities, Oracle, AZ, 1989) (Oracle, AZ: Synergetic Press, 1990), 17.
11. Allen, quoted in Parent-Altier, 44.
12. Allen, *Me and the Biospheres*, 23.
13. John Allen, "Biospheric Theory and Report on Overall Biosphere 2 Design and Performance During Mission One (1991–1993)" (paper presented at Fourth International Conference on Closed Ecological Systems: Biospherics and Life Support, Linnean Society, London, April 10, 1996), http://www.biospherics.org/linnean.html (accessed August 2008).
14. The term "Earth Jazz" comes from writer Evan Eisenberg in *The Ecology of Eden*.
15. Caravan of Dreams Theater, *Gilgamesh*, in *Collected Works*, 1:10.
16. Vladimir Vernadsky, *The Biosphere* (abridged) (Oracle, AZ: Synergetic Press, 1986), 16.
17. Vernadsky, "The Biosphere and the Noösphere," *American Scientist* 33, no. 1 (January 1945): 9.
18. Daniel Dennett, *Darwin's Dangerous Idea: Evolution and the Meanings of Life* (New York: Simon and Schuster, 1995), 62.
19. Allen and Nelson, 55.
20. Dolphin, *Tin Can Man*, 145.
21. P. D. Ouspensky, *The Psychology of Man's Possible Evolution* (London: Routledge and Kegan Paul, 1978), 10–11.
22. Allen, *Me and the Biosphere*, draft manuscript (2002), 39.
23. Bass, audiotaped interview.
24. William J. Broad, "As Biosphere Begins, Its Patron Reflects on Life," *New York Times*, September 24, 1991, B5.

25. C. S. Holling, "What Barriers? What Bridges?" in *Barriers and Bridges to the Renewal of Ecosystems and Institutions*, ed. Lance H. Gunderson, C. S. Holling, and Stephen S. Light (New York: Columbia University Press, 1995).
26. John Allen to author, September 10, 2008.
27. John Allen to author, December 12, 2006.
28. Roy Walford et al., "'Biospheric Medicine' as Viewed from the Two-Year First Closure of Biosphere 2," *Aviation, Space, and Environmental Medicine* 67, no. 7 (July 1996): 613.
29. Roy Walford, quoted in Ansley, 68.
30. Poynter, 46.

PIONEERING

1. Alling and Nelson, 2.
2. "Reflections from the Crew," *Biosphere 2 Newsletter* 2, no. 3 (1992): 9.
3. Sally Silverstone, "Ask Sally," *Biosphere 2 Newsletter* 2, no. 3 (1992): 19.
4. Henry David Thoreau, *Walden*, ed. J. Lyndon Shanley (Princeton, NJ: Princeton University Press, 1971), 90.
5. Linda Leigh, quoted in Allen, *Biosphere 2*, 31.
6. Roy Walford et al., "The Calorically Restricted Low-fat Nutrient-dense Diet in Biosphere 2 Significantly Lowers Blood Glucose, Total Leukocyte Count, Cholesterol, and Blood Pressure in Humans," *Proceedings of the National Academy of Sciences* 89 (1992): 11533.
7. Roy Walford, "Dr. Walford and Caloric Restriction," http://www.walford.com (accessed August 2008).
8. Walford et al., "Biospheric Medicine."
9. Linda Leigh et al., "Tropical Rainforest Biome of Biosphere 2: Structure, Composition and Results of the First 2 Years of Operation," *Ecological Engineering* 13 (1999): 65–93.
10. Space Biospheres Ventures, "Research and Development Report on the Status of Biosphere 2 at the End of Mission One," press release, February 10, 1994, http://www.biospheres.com/resrad.html (accessed August 2008).
11. Roy Walford, *The Voyage of Biosphere 2*, video documentary (Emergent Properties, 2005).
12. Roy Walford, "Biosphere 2 as Voyage of Discovery: The Serendipity from Inside," *BioScience* 52 (2002): 259–63.
13. Garrett Hardin, "The Tragedy of the Commons," *Science* 162 (1968): 1243–48.
14. Alling and Nelson, 8.

15. Carl Hodges, "Biosphere II—The Key Variables" (presentation at Biospheres Conference II, Oracle, AZ, September 6–9, 1985).
16. Rappaport.
17. J. K. Wetterer et al., "Ecological Dominance by *Paratrechina Longicornis* (Hymenoptera: Formicidae), an Invasive Tramp Ant, in Biosphere 2," *Florida Entomologist* 82, no. 3 (September 1999): 387.
18. John Allen, "Biospheric Theory and Report on Overall Biosphere 2 Design and Performance," *Life Support and Biosphere Science* 4, no. 3/4 (1997): 104–5.
19. Felix Gillette, "Biosphere 2: An Interview with Tony Lambard Burgess," *Land Forum* 7 (1999): 80.
20. Warshall, 22, 27.

THROUGH THE LOOKING GLASS

1. A. K. Dewdney, *Yes, We Have No Neutrons: An Eye-Opening Tour through the Twists and Turns of Bad Science* (New York: Wiley, 1998).
2. Douglas Kreutz, "Scientists Enter Research Module, New World," *Arizona Daily Star,* March 9, 1989. Emphasis added.
3. Peter Aleshire, "Biosphere Crew's 'Last Day' on Earth," *Arizona Republic,* September 26, 1991.
4. Nash, 141–42.
5. Jon Krakauer, *Into the Wild* (New York: Doubleday, 1996), 168.
6. William Cronon, "The Trouble with Wilderness," in Cronon, *Uncommon Ground,* 78–80.
7. Matt Samelson, "Biosphere Tours," *VIA,* January/February 2003, 49.
8. Marc Cooper, "Take This Terrarium and Shove It," *Village Voice* 36, no. 14 (April 2, 1991): 24.
9. Linda Leigh, "Biosphere 2 Report from the Inside," *Buzzworm: The Environmental Journal* 3, no. 6 (November/December 1991): 23.
10. Thomas Ropp, "Eco-Tourism," *Arizona Republic,* May 17, 1992.
11. David Chandler, "Biosphere Team Trying to Clear Clouds of Doubt," *Boston Globe,* April 12, 1992, 26.
12. Marc Cooper, "Debunking Biosphere," *Tucson Weekly,* November 6–12, 1991, 4.
13. Gail Tabor, "Biospherian Hurt in Accident," *Arizona Republic,* October 11, 1991.
14. Chandler.
15. Jim Erickson, "Biosphere 2's Imperfect Mission," *Arizona Daily Star,* September 19, 1993.
16. Howard T. Odum, "Scales of Ecological Engineering," *Ecological Engineering* 6 (1996): 12.

17. Gabriel Weimann, *Communicating Unreality: Modern Media and the Reconstruction of Reality* (Thousand Oaks, CA: Sage Publications, 2000), 5.
18. Cooper, "Take This Terrarium," 25. Italics in original.
19. Ruben Hernandez, "Dawn Near for New World," *Tucson Citizen*, September 25, 1991.
20. "Biosphere 2: Trying to Clear the Air," *Tucson Citizen*, July 20, 1992.
21. Andrew Ross, *Strange Weather: Culture, Science, and Technology in the Age of Limits* (New York: Verso, 1991), 23–25.
22. Odum, "Scales."
23. Unknown newscast, quoted in Walford, *The Voyage of Biosphere 2*.
24. Allen interview by John Brockman.
25. Laskas, 72.
26. Carolyn Merchant, *The Death of Nature: Women, Ecology and the Scientific Revolution* (San Francisco: Harper and Row, 1980), 290.
27. Peter Dear, *Discipline and Experience: The Mathematical Way in the Scientific Revolution* (Chicago: University of Chicago Press, 1995), 1–6.
28. Margret Augustine, "Biosphere II: Biotechnology for Earth and Space Habitats," in *Biotechnology for Aerospace Applications*, ed. John W. Obringer and Henry S. Tillinghast (The Woodlands, TX: Portfolio Publishing Company 1989), 243–56.
29. Arthur Galston, quoted in Laskas, 72.
30. Odum, "Scales," 14.
31. Peter Galison and Bruce Hevly, eds., *Big Science: The Growth of Large-Scale Research* (Stanford, CA: Stanford University Press, 1992).
32. Alling and Nelson, 14.
33. Hodges, "The Key Variables."
34. Walford, "Biosphere 2 as Voyage."
35. Howard T. Odum, "Multiple Scale Ideal in Environmental Research and Publication" (unpublished paper, University of Florida, Gainesville, August 1998), 1–2.
36. C. S. Holling, "Two Cultures of Ecology," *Conservation Ecology* 2, no. 2 (1998), http://www.ecologyandsociety.org/vo12/iss2 (accessed August 2008).
37. Mark Nelson et al., "Using a Closed Ecological system to Study Earth's Biosphere: Initial Results from Biosphere 2," *BioScience* 43 (1993): 235.
38. John Allen and Mark Nelson, "Biospherics and Biosphere 2, Mission One (1991–1993)," *Ecological Engineering* 13 (1999): 15–29.
39. Linda Leigh, "Linda's Journal—Oxygen," *Biosphere 2 Newsletter* 3, no. 1 (1993): 4–5.
40. Jeffrey Severinghaus et al., "Oxygen Loss in Biosphere 2," *EOS: Transactions of the American Geophysical Union* 75 (1994): 33–36.

41. William Knott, quoted in Victor Dricks, "Out of Thin Air," *Phoenix Gazette,* January 24, 1993.
42. Dan Sorenson, "Project's Woes May Aid Science," *Tucson Citizen,* September 27, 1993.
43. Allen, "Biospheric Theory," 104.
44. Mark Nelson to author.

THE HUMAN EXPERIMENT

1. Poynter, 208.
2. Ibid., 216.
3. Steve Yozwiak and Dave Cannella, "Crew Members Mum about Sex, Relationships," *Arizona Republic,* September 25, 1993.
4. M. Ephimia Morphew, "Psychological and Human Factors in Long-Duration Spaceflight," *McGill Journal of Medicine* 6, no. 1 (2001): 74–80, 76.
5. Poynter, 235.
6. John Sturgeon, "The Psychology of Isolation," http://scicom.ucsc.edu/SciNotes/0001/crazy.htm (accessed November 2005).
7. Poynter, 207.
8. Ibid., 251.
9. Abigail Alling, Mark Nelson, Sally Silverstone, and Mark Van Thillo, "Human Factor Observations of the Biosphere 2, 1991–1993, Closed Life Support Human Experiment and Its Application to a Long-termed Manned Mission to Mars," *Life Support and Biosphere Science* 8 (2002): 77.
10. Quoted in Walford, *The Voyage of Biosphere 2.*
11. Allen, *Me and the Biospheres,* 255.
12. Charles Kelly, "Project's Brain Staying in Background," *Arizona Republic,* September 23, 1993.
13. Bass, audiotaped interview.
14. NASA, *Spinoff* (Washington, DC: U.S. Government Printing Office, 1988).
15. John Allen to author, December 12, 2006.
16. Margret Augustine, "Mission Completed," *Biosphere 2 Newsletter* 3, no. 3 (Fall 1993): 2.

THE RESET BUTTON

1. Johnny Dolphin, *Off the Road: Poetry 1989–2000* (Santa Fe, NM: Synergetic Press, 2000).
2. Ibid., 80.

3. Steve Bannon, quoted in Dan Sorenson, "Biosphere 2 Changes Merely 'Hitting Reset,' New CEO Says," *Tucson Citizen*, March 16, 1995, 2B.
4. Wallace S. Broecker, "The Biosphere and Me," *GSA Today* 6, no. 7 (July 1996): 1.
5. Jocelyn Kaiser, "Wiping the Slate Clean at Biosphere 2," *Science* 265 (August 19, 1994): 1027.
6. Tim Appenzeller, "Biosphere 2 Makes a New Bid for Scientific Credibility," *Science* 263 (March 11, 1994): 1368.
7. Broecker, 7.
8. Jay Davis, "Environmental Technology and Engineering in Biosphere 2" (white paper presented at Biosphere 2, Oracle, AZ, August 1994).
9. Bill Harris, quoted in Anne T. Denogean, "Its Lessons Learned, Biosphere 2 Seeks Niche as Teaching Center," *Tucson Citizen*, April 17, 1998.
10. "Rehabilitating Biosphere," *Arizona Daily Star*, [c. September 1994], A14.
11. Miriam Davidson, "Arizona's Biosphere Tries to Move from Theme Park to Science Lab," *Christian Science Monitor*, January 25, 1995.
12. "Scientists Exit Biosphere 2, Ending 6 Months of Research," *Boston Globe*, September 18, 1994.
13. "Biosphere Entering Brave New World of Hard Science?" *Atlanta Journal Constitution*, September 18, 1994.
14. "Biosphere 2 a Joke No Longer," ABC News, http://www.abcnews.com (accessed 1999).
15. Steven Shapin, *A Social History of Truth: Civility and Science in Seventeenth-Century England* (Chicago: University of Chicago Press, 1994), 15.
16. Robert Merton, "The Matthew Effect in Science," in *The Sociology of Science* (Chicago: University of Chicago Press, 1973), 445.

THE NEW NEW WORLD

1. Columbia University Earth Institute, http://www.earthinstitute.columbia.edu (accessed March 2000).
2. Allen and Nelson, 55.
3. Walford, "Biosphere 2 as Voyage," 263.
4. William C. Harris, "Biosphere 2 Center Update," *EARTHmatters*, Columbia University Earth Institute, Winter 1998–99, 30.
5. Columbia University Biosphere 2 Center, *Introduction to Biosphere 2 Center as a Research Facility*, March 1, 1999, 1, 4.

6. Lynn Margulis and Gregory Hinkle, "The Biota and Gaia: One Hundred and Fifty Years of Support for Environmental Sciences," in *Slanted Truths: Essays on Gaia, Symbiosis, and Evolution*, ed. Lynn Margulis and Dorion Sagan (New York: Springer-Verlag, 1997), 219.
7. Robert M. May, "Unanswered Questions in Ecology," *Philosophical Transactions of the Royal Society of London* 354 (1999): 1951–59.
8. James Lovelock, *The Ages of Gaia: A Biography of Our Living Earth* (New York: W. W. Norton and Co., 1988), xvii.
9. National Science Foundation, "Biocomplexity in the Environment Homepage," http://www.nsf.gov/home/crssprgm/be/ (accessed March 2000).
10. For example, it has been an issue in clinical trials of AIDS treatments. See Steven Epstein, *Impure Science: AIDS, Activism, and the Politics of Knowledge* (Berkeley: University of California Press, 1996), 256.
11. Biosphere 2 Center, "Undergraduate Earth Semester Overview," http://www.bio2.edu/education/earth_sem.htm (accessed March 2000).
12. Brendan Marten, "Panel Addresses G-CEP issues," *Stanford Daily*, November 26, 2002, 5.
13. World Commission on Environment and Development, *Our Common Future: Report of the World Commission on Environment and Development* (published as annex to United Nations General Assembly document A/42/427, *Development and International Co-operation: Environment*, August 2, 1987).
14. Stephen M. Meyer, "The Role of Scientists in the 'New Politics,'" *Chronicle of Higher Education* 41, no. 37 (1995): B1–B2.
15. George W. Bush, speech announcing the Clear Skies and Global Climate Change Initiatives, Silver Spring, Maryland, February 14, 2002, "Weekly Compilation of Presidential Documents," http://www.lib.umich.edu/govdocs/text/clear.txt (accessed August 2008).
16. Environmental News Network, "Acid Rain Eats Away at Northeast," April 3, 2000, http://archives.cnn.com/2000/Nature/04/03/acid.rain.enn (accessed 2002).
17. Leslie R. Alm, *Crossing Borders, Crossing Boundaries: The Role of Scientists in the U.S. Acid Rain Debate* (Westport, CT: Praeger, 2000).
18. Leslie Roberts, "Learning from an Acid Rain Program," *Science* 251 (1991): 1302–5.
19. Walter Gilbert, "A Vision of the Grail," in *The Code of Codes: Scientific and Social Issues in the Human Genome Project*, ed. Daniel Kevles and Leroy Hood (Cambridge, MA: Harvard University Press, 1992), 84; National Research Council, *Mapping and Sequencing the Human Genome* (Washington, DC: National Academy Press, 1988), 11.
20. "Origines de la vie: le difficile de décryptage," *Le Figaro*, June 20, 1988, quoted in Daniel J. Kevles, "Out of Eugenics: The Historical Politics of the Human Genome," in Kevles and Hood, 29.

21. All Species Foundation, http://www.all-species.org (accessed 2001).
22. Richard C. Lewontin, *Biology as Ideology: The Doctrine of DNA* (New York: HarperCollins, 1991), 75–76.
23. Martin Bowen, quoted in Larry Copenhaver, "Biosphere May Be Finished as College Campus," *Tucson Citizen*, September 9, 2003.

EPILOGUE

1. Linda Leigh, "Understanding the Connections," *Biosphere 2 Newsletter* 2, no. 4 (1992): 11.
2. Allen, *Me and the Biospheres*, 116.
3. Ibid., 291.
4. Tony Burgess, "Lessons."
5. Mary Catherine Bateson, "We Are Our Own Metaphor," *Whole Earth*, Fall 1999, 14.
6. Martín Prechtel, *Secrets of the Talking Jaguar: A Mayan Shaman's Journey to the Heart of the Indigenous Soul* (New York: Tarcher Putnam, 1998), 280–81.
7. Ibid., 281.
8. Manuel and Manuel, 811.
9. Edward O. Wilson, *Consilience: The Unity of Knowledge* (New York: Knopf, 1998), 291.
10. Ken Wilber, *The Marriage of Sense and Soul: Integrating Science and Religion* (New York: Random House, 1998), 3.
11. Joseph Campbell, *The Power of Myth* (New York: Doubleday, 1988), 22–31.
12. David Williamson, quoted in Associated Press, "Landmark Biosphere 2 Threatened by Developer's Plans," February 11, 2006.
13. Fred A. Bernstein, "Sprawl Outruns Arizona's Biosphere," *New York Times*, May 28, 2006, real estate section.

SELECTED BIBLIOGRAPHY

SOURCES USED THAT RELATE DIRECTLY TO BIOSPHERE 2:

Allen, John. *Biosphere 2: The Human Experiment.* New York: Penguin Books, 1991.

———. "Biospheric Theory and Report on Overall Biosphere 2 Design and Performance." *Life Support and Biosphere Science* 4, no. 3/4 (1997): 95–108.

———. *Me and the Biospheres.* Santa Fe, NM: Synergetic Press, 2008.

———. *Succeed: A Handbook on Structuring Managerial Thought.* Oracle, AZ: Synergetic Press, 1985.

Allen, John, and Mark Nelson. *Space Biospheres.* Oracle, AZ: Synergetic Press, 1989.

Allen, John, Tango Parrish, and Mark Nelson. "The Institute of Ecotechnics: An Institute Devoted to Developing the Discipline of Relating Technosphere to Biosphere." *Environmentalist* 4 (1984).

Alling, Abigail, and Mark Nelson. *Life Under Glass: The Inside Story of Biosphere 2.* Oracle, AZ: Biosphere Press, 1993.

Alling, Abigail, Mark Nelson, Sally Silverstone, and Mark Van Thillo. "Human Factor Observations of the Biosphere 2, 1991–1993, Closed Life Support Human Experiment and Its Application to a Long-termed Manned Mission to Mars." *Life Support and Biosphere Science* 8 (2002): 71–82.

Augustine, Margret. "Biosphere II: Biotechnology for Earth and Space Habitats." In *Biotechnology for Aerospace Applications,* edited by John W. Obringer and Henry S. Tillinghast, 243–56. The Woodlands, TX: Portfolio Publishing Company, 1989.

Broecker, Wallace S. "The Biosphere and Me." *GSA Today* 6, no. 7 (July 1996): 1–7.

Caravan of Dreams Theater. *The Collected Works of Caravan of Dreams Theater.* Vol. 1. London: Synergetic Press, 1984.

Dolphin, Johnny (John Allen). *The Collected Works of Caravan of Dreams Theater.* Vol. 2. London: Synergetic Press, 1984.

———. *Liberated Space.* Santa Fe, NM: Synergetic Press, 2000.

———. *Off the Road: Poetry 1989–2000*. Santa Fe, NM: Synergetic Press, 2000.

Hawes, Phil. "Transdisciplinarity and the Biosphere 2 Project." Paper written for the First World Congress of Transdisciplinarity, Setúbal, Portugal, November 3–6, 1994.

Institute of Ecotechnics. *Man, Earth, and the Challenges*. Proceedings of the Planet Earth conference, Institute of Ecotechnics, Aix-en-Provence, France, December 12–15, 1980. Santa Fe, NM: Synergetic Press, 1981.

Kelly, Kevin. *Out of Control: The Rise of Neo-Biological Civilization*. Reading, MA: Addison-Wesley, 1994.

Marino, Bruno D. V., and Howard T. Odum, eds. *Biosphere 2: Research Past and Present*. Special issue of *Ecological Engineering* 13, nos. 1–4 (June 1999).

Nelson, Mark, Tony Burgess, Abigail Alling, Norberto Alvarez-Romo, William Dempster, Roy Walford, and John Allen. "Using a Closed Ecological System to Study Earth's Biosphere: Initial Results from Biosphere 2." *BioScience* 43 (1993): 225–36.

Nelson, Mark, and Gerald Soffen. *Biological Life Support Systems*. Proceedings of the Workshop on Biological Life Support Technologies: Commercial Opportunities, Biosphere 2, Oracle, AZ, 1989. Oracle, AZ: Synergetic Press, 1990.

Odum, Howard T. "Scales of Ecological Engineering." *Ecological Engineering* 6 (1996): 7–19.

Poynter, Jane. *The Human Experiment: Two Years and Twenty Minutes Inside Biosphere 2*. New York: Thunder's Mouth Press, 2006.

Rappaport, Irving. *Will Apples Grow on Mars?: The Glass Ark*. Video produced by Landseer Film and TV Productions Ltd., 1991.

Severinghaus, Jeffrey P., Wallace S. Broecker, William F. Dempster, Taber MacCallum, and Martin Wahlen. "Oxygen Loss in Biosphere 2." *EOS: Transactions of the American Geophysical Union* 75 (1994): 33–36.

Vernadsky, Vladimir. *The Biosphere*. Abr. ed. Oracle, AZ: Synergetic Press, 1986.

———. "The Biosphere and the Noösphere." *American Scientist* 33, no. 1 (January 1945): 1–12.

Veysey, Laurence. *The Communal Experience: Anarchist and Mystical Communities in Twentieth-Century America*. Chicago: University of Chicago Press, 1978.

Walford, Roy. "Biosphere 2 as Voyage of Discovery: The Serendipity from Inside." *BioScience* 52 (2002): 259–63.

———. *The Voyage of Biosphere 2*. Video documentary. Emergent Properties, 2005.

Walford, Roy, R. Bechtel, et al. "'Biospheric Medicine' as Viewed from the Two-Year First Closure of Biosphere 2." *Aviation, Space, and Environmental Medicine* 67, no. 7 (July 1996): 609–17.

Walford, Roy, S. B. Harris, and M. W. Gunion. "The Calorically Restricted Low-fat Nutrient-dense Diet in Biosphere 2 Significantly Lowers Blood Glucose, Total Leukocyte Count, Cholesterol, and Blood Pressure in Humans." *Proceedings of the National Academy of Sciences* 89 (1992): 11533–37.

Warshall, Peter. "Lessons from Biosphere 2: Ecodesign, Surprises, and the Humility of Gaian Thought." *Whole Earth*, Spring 1996, 22–27.

Wetterer, J. K., et al. "Ecological Dominance by *Paratrechina Longicornis* (Hymenoptera: Formicidae), an Invasive Tramp Ant, in Biosphere 2." *Florida Entomologist* 82, no. 3 (September 1999): 381–88.

SOURCES ON RELATED TOPICS THAT CONTRIBUTE TO THE CONCEPTUAL FRAMEWORK FOR THE BOOK:

Alm, Leslie R. *Crossing Borders, Crossing Boundaries: The Role of Scientists in the U.S. Acid Rain Debate*. Westport, CT: Praeger, 2000.

Campbell, Joseph. *The Power of Myth*. New York: Doubleday, 1988.

Cronon, William, ed. *Uncommon Ground: Rethinking the Human Place in Nature*. New York: W. W. Norton, 1996.

Dear, Peter. *Discipline and Experience: The Mathematical Way in the Scientific Revolution*. Chicago: University of Chicago Press, 1995.

Dennett, Daniel. *Darwin's Dangerous Idea: Evolution and the Meanings of Life*. New York: Simon and Schuster, 1995.

Eisenberg, Evan. *The Ecology of Eden*. New York: Alfred Knopf, 1998.

Gunderson, Lance H., C. S. Holling, and Stephen S. Light, eds. *Barriers and Bridges to the Renewal of Ecosystems and Institutions*. New York: Columbia University Press, 1995.

Holling, C. S. "Two Cultures of Ecology." *Conservation Ecology* 2, no. 2 (1998). http://www.ecologyandsociety.org/vol2/iss2 (accessed August 2008).

Kanter, Rosabeth Moss. *Commitment and Community: Communes and Utopias in Sociological Perspective*. Cambridge, MA: Harvard University Press, 1972.

Lovelock, James. *The Ages of Gaia: A Biography of Our Living Earth*. New York: W. W. Norton, 1988.

Manuel, Frank E., and Fritzie P. Manuel. *Utopian Thought in the Western World*. Cambridge, MA: Harvard University Press, 1979.

Margulis, Lynn, and Dorion Sagan, eds. *Slanted Truths: Essays on Gaia, Symbiosis, and Evolution*. New York: Springer-Verlag, 1997.

May, Robert M. "Unanswered Questions in Ecology." *Philosophical Transactions of the Royal Society of London* 354 (1999): 1951–59.

Merchant, Carolyn. *The Death of Nature: Women, Ecology and the Scientific Revolution*. San Francisco: Harper and Row, 1980.

———. *Reinventing Eden: The Fate of Nature in Western Culture*. New York: Routledge, 2003.

Merton, Robert. *The Sociology of Science*. Chicago: University of Chicago Press, 1973.

Meyer, Stephen M. "The Role of Scientists in the 'New Politics.'" *Chronicle of Higher Education* 41, no. 37 (1995): B1–B2.

Morphew, M. Ephimia. "Psychological and Human Factors in Long-Duration Spaceflight." *McGill Journal of Medicine* 6, no. 1 (2001): 74–80. http://www.medicine.mcgill.ca/mjm/v06n01/index.htm (accessed August 2008).

Nash, Roderick. *Wilderness and the American Mind*. 3rd ed. New Haven, CT: Yale University Press, 1982.

Pitzer, Donald E., ed. *America's Communal Utopias*. Chapel Hill: University of North Carolina Press, 1997.

Ross, Andrew. *Strange Weather: Culture, Science, and Technology in the Age of Limits*. New York: Verso, 1991.

Shapin, Steven. *A Social History of Truth: Civility and Science in Seventeenth-Century England*. Chicago: University of Chicago Press, 1994.

Weimann, Gabriel. *Communicating Unreality: Modern Media and the Reconstruction of Reality*. Thousand Oaks, CA: Sage Publications, 2000.

Worster, Donald. *The Wealth of Nature: Environmental History and the Ecological Imagination*. New York: Oxford University Press, 1993.

INDEX

Page numbers in bold font indicate illustrations.